Flávia Da Silva Bortoloti

Territorial developments

Flávia Da Silva Bortoloti

Territorial developments

The Land Credit Programme in Londrina and Tamarana - PR

ScienciaScripts

Imprint

Any brand names and product names mentioned in this book are subject to trademark, brand or patent protection and are trademarks or registered trademarks of their respective holders. The use of brand names, product names, common names, trade names, product descriptions etc. even without a particular marking in this work is in no way to be construed to mean that such names may be regarded as unrestricted in respect of trademark and brand protection legislation and could thus be used by anyone.

Cover image: www.ingimage.com

This book is a translation from the original published under ISBN 978-613-9-65188-7.

Publisher:
Sciencia Scripts
is a trademark of
Dodo Books Indian Ocean Ltd. and OmniScriptum S.R.L publishing group

120 High Road, East Finchley, London, N2 9ED, United Kingdom
Str. Armeneasca 28/1, office 1, Chisinau MD-2012, Republic of Moldova, Europe
Printed at: see last page
ISBN: 978-620-7-85465-3

Copyright © Flávia Da Silva Bortoloti
Copyright © 2024 Dodo Books Indian Ocean Ltd. and OmniScriptum S.R.L publishing group

ACKNOWLEDGEMENTS

To Antônia and Antônio, dedicated, marvellous, singular parents. Without them, nothing in my life would be possible.

To Dona Fátima, an enlightened person whom God placed in my path, providing support even from far away.

To all the families I visited, because without them this work would be unrealisable. For welcoming me and sharing their reality with me.

To Andréia and Karen, beloved sisters who always encouraged me.

To Luiz Felipe, my beloved godson who, even from afar, brings happiness to my days.

To Eliane, more than a supervisor, my "intellectual mother". She always believed in and encouraged my work.

To Rafael, without whom the field surveys would not have been possible. For his shoulder to lean on, for believing in me when all seemed lost.

To my aunts Vera and Cida for their unconditional encouragement.

To Ricardo Venturelli, for everything, for listening to me complain, for complaining with me, for his company, for the maps and materials developed for this work.

To Carlos Bortolo, dear friend, for all the endless conversations.

To Vanessa, a friend who always brought new ideas and for her unconditional friendship.

To Maria Abadia (Bá), for her friendship, advice and for believing that I could make a difference in a positive way.

To the EMATER technicians, Paulo Mrtvi and Marcelo Campos, who, as well as providing valuable information, supplied many of the photos and maps that appear in this work.

To Marcio Silva of the Secretariat of Agriculture and Supply of Paraná - Curitiba

To my fellow Masters students.

To the professors of the examining committee.

To the teachers in the Geography Department who helped me from the start of my degree and encouraged me to go further.

To the Araucária Foundation for financial support in 2009/2010.

"Land ownership presupposes that certain people have the monopoly of disposing of certain portions of the globe as private spheres of their particular will, to the exclusion of all other wills" (Karl Marx).

SUMMARY

The aim of this research is to analyse the Brazilian agrarian question, considering everything from the constitution of the latifundia to the new face of agrarian reform, adopted by the Brazilian state but guided by the neoliberal ideals of international financial organisations such as the World Bank, which are materialised in space through the Land Credit Programmes. We therefore propose to examine the way in which land reorganisation takes place, through the actions undertaken by agents of national and international capital with the backing of the Brazilian state; They act on the land structure not with the intention of changing it, but rather to make it even more concentrated and inaccessible to thousands of workers in the struggle for land, always acting where there is tension, in an attempt to destructure social movements and historical struggles for access to land, thus transforming agrarian reform from a social and economic problem into a market issue, through programmes that favour the purchase and financing of land, to the detriment of expropriation and redistribution. In this paper, we will also examine the contradictory re-creation of the peasantry within the capitalist mode of production, using the Banco da Terra land credit programme as a parameter. We are combining two scales of analysis here: one that focuses on the general implications of the Brazilian agrarian question and the other on this public intervention programme in the country's land system, in order to position the debate from an empirically based investigation: the Banco da Terra groups in Londrina and Tamarana.

Keywords: Land credit programmes. Peasant recreation.

Land redevelopment. Land bank.

INTRODUCTION

Since Brazil's entry into the mechanism of unequal exchange as an exploitative colony and the donation of hereditary captaincies and sesmarias, Brazil's land ownership structure has had a strong exclusionary and concentrating character, which is why we sought to discuss how this characteristic became more intense after the adoption of policies advocated by the World Bank (WB) to resolve land conflicts in countries classified as peripheral.

This discussion is based on an analysis of the Banco da Terra land credit programme, set up by Complementary Law No. 93 of 1998 and Decree No. 3475 of 2000, which was abolished in February 2003. However, only the nomenclature no longer exists, as the programme is still in operation under the name of the National Land Credit Plan (PNCF), thus imposing a new dynamic on the land market in Brazil. We therefore sought to ascertain how these processes materialise in Londrina and Tamarana.

The Brazilian agrarian question has been discussed for many years, not just in the field of geography. However, there is a lack of specific studies on land policies, particularly those instituted as part of land credit. This work is therefore justified by the relevance of collecting consistent data on ongoing projects, and by the possibility of contributing, from this study, with parameters for public action in land management in the country.

We highlight the importance of discussing the ways in which the market-driven land redevelopment model has been implemented, to the detriment of agrarian reform, i.e. land expropriation that does not fulfil its social function. This research is also characterised by problematising the territorial conflicts at the heart of the struggle for agrarian reform and by deepening our understanding of the agrarian question in Brazil.

It is also important to investigate the contradictory re-creation of the peasantry and the obstacles to maintaining peasant logic in the face of impositions from the land credit project. We thus believe we can contribute to the formulation of parameters for action by movements in the struggle for land, as well as to promoting alternative public policies to those implemented in the cases in question.

Finally, this work aims to understand the mediating role of the state in the issue of private land ownership and to understand the meaning of mediation by global financial organisations in the agrarian question in peripheral countries.

The choice of this topic began with a paper published at the 3rd

International Symposium on Agrarian Geography, held in Londrina in 2007, which eventually led to us joining the Scientific Initiation Programme at the State University of Londrina, where we carried out studies on the King of Lettuce Group.

Our research methodology used secondary data from institutions such as the Brazilian Institute of Rural Assistance and Extension (EMATER), the Ministry of Agrarian Development (MDA), the Paraná Secretary of Agriculture and Supply (SEAB), the Department of Rural Economy (DERAL) and the National Institute for Colonisation and Agrarian Reform (INCRA), as well as primary data collection through fieldwork in all the Banco da Terra projects in the municipalities of Londrina and Tamarana, as shown in Figure 1.

Primary data collection took place between February and July 2010, through visits and interviews guided by the questionnaire presented in Appendix I. There are a total of 210 families in the eight projects studied, as shown in Table 2 and Table 3; we chose half of these families to interview as the universe of analysis.

In the first chapter, "The Brazilian agrarian question: from the constitution of the latifundia to the land credit policy", we address the more general issues of the Brazilian land structure, pointing out how it was dealt with in colonial Brazil, in the Old Republic, in the New Republic, in the military government and, finally, in the neoliberal administrations of the governments after 1990.

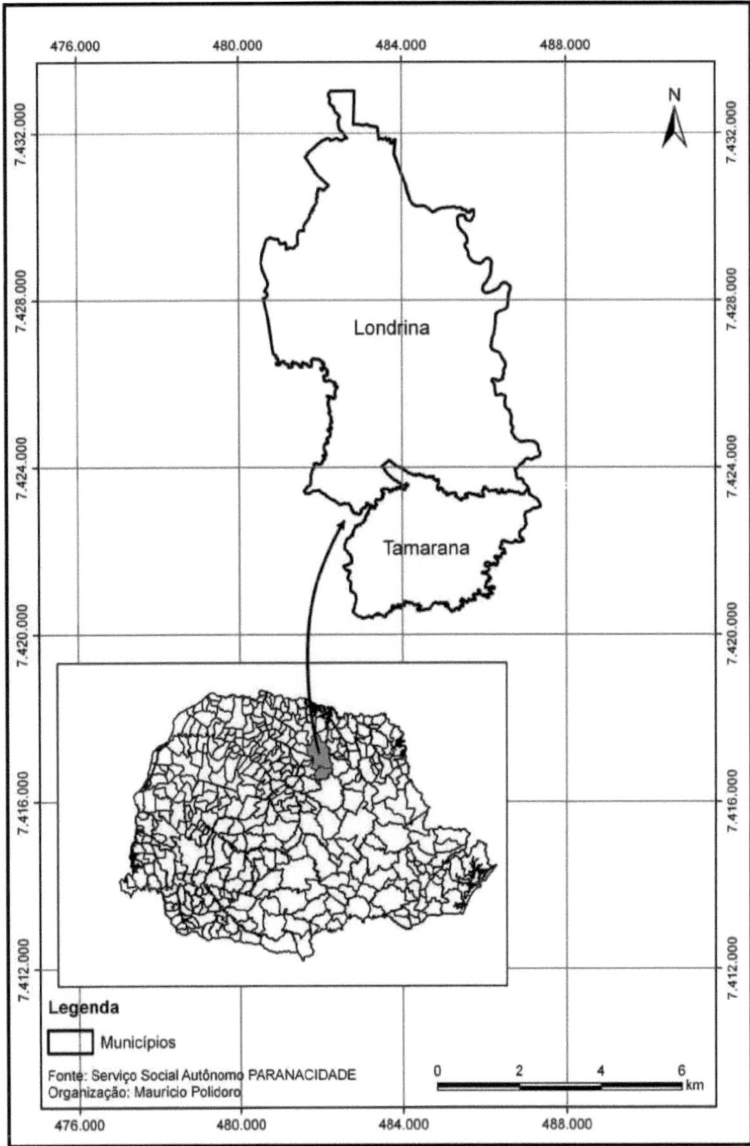

Figure 1 - Geographical location of the municipalities of Londrina and Tamarana

The second chapter, "The implementation and operation of land credit programmes in Brazil", studies the operation of the Cédula da Terra (Land Credit) programmes, a pilot project that covered the states of Ceará, Maranhão, Pernambuco, Bahia and the northern region of Minas Gerais; the Banco da Terra (Land Bank), which was extended to all the

7

states of the federation; and the National Land Credit Plan (PNCF). We discuss the way in which these programmes have been used as an agrarian reform policy, to the detriment of state-led agrarian reform, based on the expropriation of land that does not fulfil its social function; we also address the way in which the conduct and materialisation of this land credit policy have contributed to the de-ideologisation of the historic struggle for the right to remain on the land.

The third chapter, "The development of capitalism in the countryside: from theories to reality", is based on the empirical scale of the Banco da Terra Programme, in which we examine the internal land organisation of the selected projects, taking into account the size of the plots, the forms of land use, and labour relations; in this part of the work, we analyse internal production in the light of two elements: production for self-consumption and production for the market.

We investigated the structures that indicate the process of peasant re-creation, such as kinship relations, mutual aid practices, community activities, among others. Finally, we looked at the financial situation of the projects, taking into account the individual capacity to pay the annual debts relating to the payment of the plot and the credits for funding and investment.

CHAPTER 1

THE DEVELOPMENT OF CAPITALISM IN THE COUNTRYSIDE: FROM THEORIES TO REALITY

Capitalism, according to classical Marxism, is the system of production that aims to obtain and maximise profits. To this end, it adopts as its basic premise the separation of the working man from the means of production, making him "free" for wage labour, which, through the extraction of surplus value (unpaid labour), materialises the capitalist's profit.

> [...] the workers become free workers, that is, freed from all ownership other than ownership of their labour power, their ability to work. [...] They also become free in the sense that they are not subjugated by anyone, by a landowner or a slave master. As well as being free, they are therefore equal to those who own property (MARTINS, 1986, p. 152).

In capitalist society, the condition of person arises from relations of exchange. On the one hand, the worker is provided with the only commodity he has to offer, labour power; and on the other, the capitalist, who owns the means of production and needs the labour power of another to produce goods. Therefore, as Martins (1986, p. 153) points out, "one person only exists through another".

> [...] to enter into the relationship of exchange, each person has to be each person, individualised, free and equal to all the others; at the same time, each person is never each person, because the existence of the person depends totally on all the other persons, on the relationships that each person establishes with the others. Each person is created in the person of the other (MARTINS, p. 153).

As Martins (1986) teaches, only labour creates value, and this value is measured by the number of hours of work socially necessary to produce the commodity. The value of labour power is measured by the part of the value created by the worker with his labour, which returns to him in the form of wages. Wages are the money paid to workers so that they can buy the things and services they need on the market to reproduce themselves as workers.

> As he no longer owns anything that would allow him to turn his labour capacity into an instrument of autonomy, of another type of freedom, like that of the craftsman or the farmer who works for himself, the salary he receives is conceived as the equivalent of what he needs: the means of life necessary for his social reproduction (MARTINS, 1986, p. 155, emphasis added).

Wages are not determined by the particular will of the individual capitalist, but by the socially determined rate of profit of capital. Thus, a capitalist cannot pay wages that are too high, because his rate of profit would fall below the average profits made by the other capitalists. Unless he is able to recover his profit over a certain period of time, this capitalist will start to make losses and, in a limit situation, will have to stop being a capitalist.

So,

> Capital is not at the service of the capitalist. On the contrary, it is the capitalist who is at the service of capital. Capital is the thing that dominates the person, not only the worker, but also the capitalist. Only then the worker loses and the capitalist wins (MARTINS, 1986, p. 154).

The capitalist social relationship presupposes relationships between legally equal people (a concept discussed in Chapter I), while at the same time producing economic results that are profoundly unequal to each other, wages and profit, personified by unequal people, the worker and the capitalist. This is only possible to the extent that labour appears to be the property of capital and not of the worker, making them alienated from labour.

> The illusion of equality and equivalence that permeates this relationship of unequal exchange, making what is the product of labour appear as the product of capital, makes the worker face the wealth that he himself produces, and which grows in the form of capital, as if it were alien to him, alienated from him. [...] he sees himself not as he is, but as he appears to be, as equal and free; not as if capital depended on him, or on his labour, but as if he depended on capital. He becomes a stranger to his work, to his labour. [...] This is what is meant when we talk about the alienation of the worker in capitalist society. He doesn't appear as the creator of wealth, of capital, but as a creature of capital itself (MARTINS, 1986, p. 156-157).

So far we have explained the bases of capitalist relations of production in order to understand the development of capitalism in the countryside. We must therefore clearly state that capitalist relations are social relations that presuppose an exchange between capital and labour. To this end, the separation

between the worker and the means of production. Capital must therefore be understood as the product of unpaid labour, of surplus value, of the conversion of labour into capital, which exceeds that which has materialised in wages.

The development of capitalism doesn't work in the same way in industry and agriculture, because differences arise from the unequal behaviour of working time and production time in these productive sectors. This can only be understood on the basis of the labour theory of value, which considers that the exchange value of any commodity, whether it is produced in a typically capitalist form or not, is determined by the labour time socially necessary to produce it (PAULINO; ALMEIDA, 2010).

> Labour time is always production time, which is the same as saying that it is a time during which capital is firmly held in waiting for production. But the opposite is not true, i.e. not all the time in which capital is kept in the production process is necessarily labour time. Production time consists of two parts, a period in which labour is actually applied to production, and a second, during which the 'unfinished' commodity must await the influence of natural processes, without simultaneously submitting to the labour process (MARX, 1974, p. 242).

Paulino and Almeida (2010, p. 31) state that:

> [...] value, and therefore surplus value, is not equal to the time that the production phase lasts, but coincides with the materialised and living labour time employed during the exact phase of production. By analytically separating working time from 'unproductive' production time, we can see that the more production time and working time coincide, the greater the productivity and self-expansion of capital. In this way, typically capitalist agriculture expands in sectors of activity where production time can be successfully reduced, and sectors that are naturally dependent on this time gap are ruled out. In turn, the opposite tends to repel capitalist investment, the gaps that the peasant class takes advantage of to recreate itself.

Through artificial processes, industry manages to equalise production time and working time. In agriculture, however, this is not the case, because in addition to human energy, natural cycles with variable and rigorously independent labour times are required to carry out production.

According to Paulino and Almeida (2010, p. 31)

> [...] an appreciable interval separates sowing and harvesting, during which the increase in value is nil, because if some interventions can shorten the cycle of germination, development and maturation, they cannot eliminate it completely, and the same applies to animal production.

We know that merchandise only gains value when it enters the circulation process, i.e. the market. Consequently,

> [...] the shorter the time in which vital energy is consumed to produce a good, coupled with the smaller amount of money immobilised to do so, the greater the potential amount of surplus value to be extracted, because capitalists can even pay wages using the value created by the workers themselves, already converted into money in the circulation process. [...] On the other hand, the more perishable a commodity is and the greater the restriction on its circulation time due to its natural characteristics, the less suitable it is for capitalist production. [...] Consequently, the capitalist unit aimed at quick returns and the creation of value cannot find the ideal

As we have already discussed, capital is the product of labour and, in order to understand the unfolding of capitalism in agriculture, we need to highlight the contradiction between land and capital. Land is a natural asset, finite and impossible to reproduce. In this way, land has no value because it is not produced by labour, nor can it be understood as capital, since it is not a product of capital. In the same way that capital appropriates labour (by monopolising the means of production), it can appropriate land, it can make land, which is neither a product of labour nor of capital, appear dominated by it. Just as the capitalist has to pay a wage to appropriate the labour power of the worker, he also has to pay a rent to appropriate the land.

According to Marx (1977, p. 714), land rent "[...] takes the form of a certain sum that the owner of the soil receives annually for renting a piece of the globe." Thus, land rent is the licence for the capitalist exploitation of land.

The meaning of the contradiction between land and capital lies in the payment of land rent. The moment the capitalist pays for its use, he is unproductively immobilising this part of his capital, exclusively for the purpose of

removing the obstacle that land ownership represents in capitalism, the reproduction of the capitalist mode in agriculture.

The payment of land rent, according to Martins (1986, p. 162):

> [...] it therefore represents irrationality for capital. This does not mean, however, that the appropriation of land by capital prevents it from being used according to capitalist criteria. The subordination of land ownership to capital occurs precisely so that it produces under the domination and according to the assumptions of capital. The capitalist appropriation of land allows the labour that takes place on it, agricultural labour, to become subordinate to capital.

Land, producing under capitalist criteria, appropriated as if it were capital, becomes the equivalent of capital. However, as much as land appears as capital, this does not make it effectively capital, because what it produces, from a capitalist point of view, is different from what capital produces. Capital produces profit through the extraction of surplus value, labour produces wages and land produces income (MARTINS, 1986).

Unlike profit, which comes from production, since the surplus value is extracted from the worker by capital in the production process, land rent arises from the distribution of surplus value.

As Martins (1986, p. 162-163) had already warned, it is in the process of production that the worker produces his wage and the capitalist extracts his profit. "When he [the capitalist] pays the rent to the owner, he is not producing anything; he is distributing part of the surplus value he has extracted from his workers."

Land rent is a tax paid by society as a whole, paid by "everyone who needs to live, feed, clothe and live, because all this requires land" (PAULINO; ALMEIDA, 2010, p. 83).

According to Martins

> Although capital pays a rent for the use of the land, in fact the landowner also shares with the owner of capital the plunder that he practised alone against the workers. [...] It [the land rent] first appears in the hands of the capitalist as if it were an extraordinary profit, which he does not think he has the right to retain for himself because for him profit is the payment for the ownership of the instruments of production, proportional to the value that these means have. He keeps his share and passes on his share to the landowner (MARTINS, 1995, pp.163-164).

However, capitalist and landowner, two objectively separate and opposed subjects, can appear united by their common interest in appropriating the surplus value produced by the workers, and can also appear unified in a single figure, that of the landowner who is also the owner of capital.

When the capitalist buys the land, he is not necessarily interested in the land itself, but in the rent from the land; he is actually buying the right to seize part of the social surplus value.

> When the capitalist buys the land, he converts his capital into capitalised income, anticipated income, the right to extract an income from the land and at the same time the right to recover his capital in full and even with an increase by simply converting capitalised income into capital (MARTINS, 1986, p. 167).

The conversion of capitalised income into capital, to which the author refers, takes place through the sale of land, either in lots or in its entirety, a situation in which the land rent is collected all at once, or by leasing parts of the property to small producers.

Thus, in the capitalist mode of production, land is not only an instrument of production, it also acts as a store of value and a guarantee for obtaining credit.

However, the development of capitalism in the countryside cannot be understood through a linear reading, believing that capitalism would act in the countryside in the same homogeneous way as it did in industry.

We analysed the Brazilian countryside, believing that the development of capitalism is contradictory, unequal (Oliveira, 1991) and permeated by heterogeneity. Contradictory, because at the same time as it extracts profit via wage labour, it creates and recreates non-capitalist labour relations (partnership, peasant family labour).

Therefore,

> [...] capital does not absolutely expand wage labour, its typical labour relationship, to every corner and place, totally and absolutely destroying peasant family labour. On the contrary, capital creates and recreates it so that its production is possible and, with it, new capitalists can also be created (OLIVEIRA, 1991, p. 20).

> [...] the development of capitalism must be understood as a (contradictory) process of expanded capitalist reproduction of capital. And this as the reproduction of non-capitalist social forms, even though the logic, the dynamic, is fully capitalist; in this sense, capitalism is nourished by non-capitalist realities, and these inequalities do not appear as historical incapacities to overcome them, but show the conditions recreated by capitalist development. In other words, the expansion of the capitalist mode of production (in its expanded capitalist reproduction of capital), as well as redefining old relations by subordinating them to its production, engenders non-capitalist relations that are equally and contradictorily necessary for its reproduction (OLIVEIRA, 1981, apud PAULINO; ALMEIDA, 2010, p. 29).

As a result, there are two distinct logics in agriculture: peasant and business. The logic of peasant agriculture is based on self-employment, family reproduction and working land, while the logic of corporate agriculture is based on the reproduction of capital, wage labour and business land. These logics lead to different forms of land appropriation and (re)production of space. "Peasant life is not organised by the needs of the market, as the capitalist unit is; it is a mode of social existence made possible by a mode of production" (PAULINO; ALMEIDA, 2010, p. 19).

Peasant production in the countryside is limited to survival and not the pursuit of average profit. In peasant labour, part of the production goes into the direct consumption of the members of the production unit and the surplus is commercialised in the form of merchandise. As expressed by Oliveira (1995), there is only the simple circulation of goods (M-D-M), in which the goods produced are commercialised with the aim of converting them into money to obtain other goods needed to satisfy needs.

Capitalist production is based on making a profit and for this to happen, the circulation of capital needs to take place in its two forms: the simple circulation of capital (D-M-D) and the expanded circulation of capital (D-M-D'), in which the initial money (achieved through the commercialisation of commodities) is used to produce other commodities which, when commercialised, will bring more money to the capitalist than was immobilised at the start of production, generating a profit.

We are faced with two distinct production logics that seem unable to coexist, because there is the idea that the establishment of the capitalist mode of production would lead to the destruction of the peasantry, as put forward by Lenin (1980, 1982), Kautsky (1980), among other theorists. However, peasant production survives within the capitalist mode of production, and this is due, as already pointed out, to the fact that capitalism uses non-capitalist relations of production to reproduce itself.

Like this,

> [...] it is very important to distinguish between the production of capital and the capitalist reproduction of capital. The production of capital is never capitalist, it is never the product of capitalist relations of production, based therefore on capital and wage labour. Therefore, not only can non-capitalist relations of production be dominated and reproduced by capital, as is the case with peasant family property, but certain relations may not appear to be part of the capital process, although they are, as is the case with capitalist land ownership (MARTINS, 1995, p. 170-171).

We should also point out that this contradictory development of capitalism stems from the fact that the production of capital never derives from specifically capitalist relations of production, based on wage labour and capital. Thus, according to Oliveira (1995, p. 11), "for the capitalist relationship to occur, its two central elements must be constituted: the capital produced and the workers deprived of the means of production".

We must therefore understand that the recreation of relations that are not essentially capitalist, contained in the contradictory development of capital, is a kind of permanent primitive accumulation of capital, and is therefore necessary for its development.

Oliveira (1995, p.11) states that "the peasantry must therefore be understood as the social class that it is. It should be studied as a labourer created by capitalist expansion [...]".

Theorists who discuss the creation and re-creation of non-essentially capitalist relations within the capitalist mode of production, such as peasant relations, point to three predominant ways in which this occurs (OLIVEIRA, 1990, p.42).

The first concerns landowners who speculate on land as a commodity; by selling it off, mainly through allotments and agricultural colonisation, they end up, contradictorily, creating the conditions for the recreation of the peasant-owner.

The second refers to the actions of the state, both as a "distributor" of land in colonisation projects and land restructuring, and as a manager of public funds, setting minimum agricultural prices, which in the end guarantee contradictory conditions for peasants to reproduce themselves. We would like to highlight here the various land credits created by the state, together with international financial institutions, so that this recreation can take place; however, we must consider that only rare exceptions manage to reproduce themselves as peasants under land credit programmes.

Finally, we should highlight the formation of cooperatives in the countryside, which originated in the 18th century as an instrument of defence for farmers against the merchants who, by acting as buyers and users, exploited the peasants, leading to their proletarianisation. Cooperatives thus came into existence in the countryside, operating in the credit and marketing sectors.

In short, situations like these can, contradictorily, strengthen and consolidate the peasant, thus allowing him to reproduce as a social being.

It is in this context of the contradictory re-creation of the peasantry within the capitalist mode of production, with the state and international financial institutions such as the World Bank (WB) and the Inter-American Bank for Reconstruction and Development, through projects such as the Land Bank, as the agents of re-creation, that we will conduct our reflections.

According to Paulino (2009, p. 69):

> [...] some of the territorial disputes in the countryside are not just about technical issues, about production itself, on which the spotlight falls. Rather, they precede it, because they comprise the struggle for access to and permanence on the land. Many of those who engage in these struggles yearn for a differentiated insertion into the world, which can be summed up in the principle of autonomy over one's own work, a fundamental element governing peasant time and space (emphasis added).

The above quote therefore shows that this is a logic that is the opposite of corporate agriculture, in which the time of capital imposes/supposes alienated labour, in which the workers do not recognise themselves in the final commodity, and do not become "participants in the results materialised in the wealth that comes from it" (PAULINO, 2009a, p. 69).

However, this does not mean that peasants are exempt from the looting perpetrated by the capitalists, but rather that their way of life has a potential that is not employed by proletarianised workers, while the latter can only reproduce themselves through the relationship of selling the only commodity they possess, labour power, "peasants are the only class that can reproduce itself independently of the others"

(PAULINO, 2009a, p. 69).

> It's clear that this is a potentiality and only applies in limited situations, because if the peasants have not been captured in the sense of real subjection or capital, in other words, if they still retain control over the means of production, they are burdened with the subjection of the land rent to capital, which means that it is not their direct labour, but the fruit of it, that forms part of the circuit of capitalist accumulation (PAULINO, 2009a, p. 69).

When it is the fruit of peasant labour that is integrated into the capital accumulation circuit, we are faced with the process referred to by Oliveira (2002) as the monopolisation of territory by capital. This process involves peasants who maintain control over the means of production and use their own labour to make them produce. It is therefore a form of agriculture in which capital promotes the subjection of peasant income from the land by controlling the intermediate stages between producer and final consumer.

> The process of monopolising the territory imposes overwork on peasants on a scale that varies according to the product, the situation and their ability to resist the plunder, because some of the agents in the industrial, commercial and financial sectors will take part of the wealth contained in the food and other goods placed on the market by the former, by manipulating prices in the intermediate circuits between producers and final consumers (PAULINO, 2009a, p. 70).

By affirming that we understand the Brazilian countryside through the thesis of the contradictory and unequal development of capitalism, we are also affirming that the peasant does not disappear within this mode of production and is contradictorily recreated by it. In this way, we do not agree with the classic thesis of the disappearance of the peasant, just as we do not agree with the trivialisation of the debate on the historical weight, which is therefore irremovable, of the colonial past in the constitution of private property on landowning bases. This is the reason for a historical dialogue with the appropriation of land and the procedural constitution of the land monopoly in the country, on which, supposedly, the land credit policy we analysed could intervene, attenuating it.

1.1 The BRAZILIAN AGRARIAN QUESTION: FROM THE CONSTITUTION OF THE LATIFUNDIUM TO THE LAND CREDIT POLICY

When dealing with the constitution of the latifundia, as well as the private ownership of land in Brazil, we must first emphasise the genesis of this Portuguese colony.

Brazil emerged as a Portuguese possession in the context of the

European Maritime Expansion of the 15th and 16th centuries, in other words, the colonialist expression of mercantilist expansion.

According to Paulino (2006, p. 66), "colonialism favoured a hitherto unprecedented transfer of intercontinental wealth, placing from the outset the nascent European nation states in a privileged position in the mechanism of unequal exchanges."

In this mechanism of unequal exchange to which the author refers, on the one hand there are European countries, capitalists/mercantilists eager for territories to conquer/dominate that could supply them with precious metals or any raw material of interest to trade, i.e. accumulation. On the other hand, territories occupied by peoples organised on a different social basis to those known to Europeans. People who, according to European ideas, could be enslaved/tanned or decimated so that the fruits of the land could be harvested and generate profit for the colonising country.

> With the invasion of Europeans, the organisation of production and the appropriation of nature's goods here came under the aegis of the laws of mercantile capitalism that characterised the historical period that was already dominant in Europe. Everything was transformed into merchandise. All productive and extractive activities were aimed at profit. And everything was sent to the European metropolis as a way of realising and accumulating capital (STEDILE, 2005, p. 20).

Brazil entered this order as an exploitative colony, in which all productive activities were geared towards accumulation in favour of the metropolis. At the first moment of colonisation, land was merely an intermediary instrument that ensured the accumulation of the metropolis, since at that time the foundation of the economy was based on slavery.

Martins (1995, p. 37) points out that

> [...] on the one hand, the foundation of slave labour lay in the slave trade; it was in the slave trade and not on the slave farm that slavery was recreated. On the other hand, this situation made sense, since it allowed slave traders to make capitalised income from the captive, to extract income from the colony even before colonial production, instead of extracting it through monopoly and territorial income.

As a result, there were two reasons why the slave trade linked the viability of the agro-export economy to the concentration of land: firstly, the scale of the trade per se and the need to block any experience of competition, such as production using free labour, given the high cost of captive labour.

Martins (1990, p.16) states that:

> The exploitation of the slave in the production process is therefore already preceded by parameters and commercial relations that determine it. This

> exploitation not only includes the average profit, but also the conversion of capital into capitalised income, the portion of the surplus that the slave can produce and which is paid in advance to the slave trader.

Unlike free workers, slaves entered the labour process stripped of any property, so they did not behave as sellers of the commodity of labour power (as free workers who were legally equal to their masters did). The captive entered this process as a commodity, but did not behave as capital, but rather as its equivalent, capitalised income, since his exploitation was determined, as Martins (1990, p. 15) states, "by the alternative use of the capital invested in him in advance [...]".

Paulino (2006, p. 66) puts it this way:

> Because they represent capitalised income, slaves are the masters' most valuable asset, which does not dispense with the need to control the land, the real means of production. As a result, it remains hostage to this group, whose prestige already ensures that it is granted letters of sesmaria and, at the same time, this status grants it inalienable powers in the political-administrative sphere (emphasis added).

By highlighting the quote above, we want to make it clear that the master's monopoly was not in the land, "the real means of production", but in the labour of the captive; property was therefore personified in the slave. Only with the introduction of free labour could ownership be transferred to the means of production; with this, land became capitalised income.

The first legal system for Brazilian land, the system of granting sesmarias, based on donations of land to Portuguese noblemen who were willing to produce and exploit in the Colony along the lines proposed by the Crown, contributed greatly to the formation of the latifundia. The fact that there were no measurements or inspections of these donations led to the formation of large farms, which suited the type of colonial exploitation adopted by the Portuguese Crown in its new possession.

Based on Silva's (2008) observations, we believe that these concessions resulted from a transposition of a legal institute existing in Portugal to the colony. In this way, territorial appropriation in Brazil was determined by two historical factors: being part of the European commercial expansion of the 15th and 16th centuries and being a Portuguese possession.

So,

> The first aspect led to the characteristics of the economic utilisation of the newly discovered lands. The second aspect determined the status of colonial soil, i.e. the transposition of the rules governing land ownership in Portugal to the new territory (SILVA, 2008, p. 25).

The legal institute we referred to earlier, the system of granting sesmarias, was instituted in Portugal at the end of the 14th century in order to solve the supply crisis. The aim was to put an end to the idleness of land not cultivated by the landlords.

Thus, according to Silva (1996, p. 37):

> Landlords who did not cultivate or lease their land lost the right to it, and the vacant land (returned to the original lord, to the Crown) was distributed to others so that they could plough it and make use of it, thus respecting the collective interest.

The sesmarias regime in Portugal was successful for what it was intended for

However, when it was transferred to the Colony, no thought was given to adapting the law to the reality of this territory of incomparably greater proportions than the Metropolis. In Portugal at the end of the 14th century, the practice of sesmarialism generally led to smallholdings. However, in Brazil, it was one of the main causes of latifundia. Thus, according to Silva (1996, p. 74), "the sesmarias system undoubtedly contributed to the formation of the colonial latifundia, insofar as it adapted to the imperatives of the colonisation system".

According to Costa Porto (apud SILVA, 2008, p. 43):

> One of the main distortions of our sesmarialism - largely the result of the inability to ignore the peculiarities of the conquest, applying to it the discipline imagined for the Metropolis - would occur with respect to the structure of land ownership, and its synthesis would be this: while in Portugal at the end of the 14th century, the practice of sesmarialism generated, as a rule, small property, in Brazil it was the main cause of latifundia.

In order for the sesmarias system in Brazil to be profitable for the Metropolis, the land had to be handed over to whoever wanted to produce it along the lines imposed by the Metropolis, such as the monoculture exploitation of sugar cane, justified by the experience of production in other Portuguese colonies, Dutch financing and the high price of the product on the metropolitan market. Another point to emphasise was the system of metropolitan "exclusivity", i.e. the monopoly of the Portuguese metropolis over the purchase of sugar and the sale of European products needed by the colony.

And finally, the fundamental characteristic for the functioning of the colonial system was the use of slave labour, both indigenous and African, because if paid labour was employed, access to land would not be totally blocked, and this blockage was essential for the plantation production model[1] to be successful in Brazilian lands. Silva (2008, p.31) puts it this way:

> The economic and social organisation of the colonies therefore had to obey the injunctions of the modern colonisation process, which pushed them towards the production of colonial surpluses. This aspect is essential for explaining one of the fundamental characteristics of the society that was organised in Brazilian lands and which indelibly marked the life of the Colony and the future independent country, namely slave labour [...].
> [...] the transfer of free labourers to the new territories would make it impossible for the mechanisms of the colonial system to function because the availability of unappropriated land would quickly turn them into owners producing for their own consumption.

Still on the subject, Silva (2008) points out that the introduction of compulsory labour made it possible, for a longer period of time, to maintain the availability of land for the social stratum that produced according to the demands of the colonial system. The maintenance of slave labour made the control of land possible and necessary, since it basically divided society between masters and slaves and because it quickly exhausted the soil. These elements contributed to Colonial Brazil becoming a highly concentrated economic and social organisation.

Even though the Crown bequeathed the process of occupying Brazilian lands to private individuals, it did not pass on the entire expanse of cultivated land to the private patrimony: only ten leagues of land could be taken wherever they wished, as long as they were not contiguous. Therefore, what was donated by the king was not the ownership of the land, as this continued to belong to the Crown, but the benefits of the usufruct of the land.

Martins (1999, p. 23) puts it this way:

> With regard to land in particular, the centre and basis of power to this day, and even more so in the colonial period, the king always maintained eminent ownership of the land granted in sesmaria. This meant preserving his right to recover possession of lands that were abandoned or not used in a way that produced the taxes to which he was entitled.

In this context of colonisation, the Metropolis did not have enough resources to provide the political-administrative apparatus in the colony, nor did it have enough to ensure the infrastructure and services that would support the economic

---

[1] Type of agricultural system based on monoculture production for export, using large estates and slave labour.

enterprise it aspired to.

For this reason, the Crown also resorted to the property of private individuals to carry out public services, giving local power and honours in exchange, stressing that the fruit of this exchange could be converted into wealth, land or money.

Martins (1999, p. 22) points out that:

> It was the private individuals who made the expeditions to fight the Indians, who built the bridges and roads, who organised and administered the villages, who waged war on the invaders. Always at the cost of their patrimony, as a political rather than economic tribute to the Crown.

On the political and economic dimensions inherent to the process in On this issue, Martins (1999, p. 23) explains that "political loyalty was rewarded with material rewards, but also with honours, such as titles and privileges, which ultimately resulted in political power and, consequently, economic power (emphasis added)."

For this reason, in Brazil a distinction was never made between the public and the private, as a distinction of rights relating to citizens. Martins (1999, p. 22) points out that:

> [...] it was a distinction that remained limited to public and private property. Therefore, it was a distinction relating to property rights and not to personal rights. Even then, it was a distinction that never gained clarity or clear contours. [...] The great distinction was of a different nature and overlapped with all the others: what was the patrimony of the king and the Crown and what was the patrimony of the municipalities, that is, of the people. And then the very concept of a person, as we know, was limited to whites and Catholics, pure of blood and pure of faith. The impure, i.e. mestizos, indigenous slaves, black slaves, as well as Moors and Jews, were subject to a gradation of exclusion that went from the condition of lord of patrimony to the condition of patrimony of lord (emphasis added).

However, the sesmarias regime cannot be considered solely responsible for the concentrated nature of Brazil's land ownership structure, due to another form of land appropriation, tenure, which in the early years of colonisation was characterised as the form of occupation of farmers unable to request sesmarias, who exploited the land based on family labour and activities aimed at self-consumption and the production of surpluses for the market. However, another type of tenure coexisted, taking the form of large estates.

The lack of control by the Crown and the non-compliance with the legal requirements of the landlord over the measurements and demarcations of sesmarias and possessions meant that during the 18th century there was disagreement between the sesmeiros, posseiros and the Crown.

The Portuguese monarchy tried to regain control of the process of territorial appropriation by making demands and threatening to suppress concessions, forums, laws and royal charters. On the other hand, appropriation through possession challenged colonial authority. The contradictions between the Colony's rural landlords and the Metropolis over the issue of territorial appropriation worsened, culminating in the end of sesmarias on 17 July 1822, which contributed significantly to the definitive break in colonial ties.

The period between the suspension of sesmarias until 18 September 1850, when the Land Law was passed, was characterised by constant occupations due to the absence of laws regulating land use and ownership. This led to an increase in the size of sugar mills and coffee plantations through ownership. The expansion of rural properties, coupled with the need to obtain labour for large-scale farming, contributed to the dynamic sectors of the economy, in the form of the owners of large coffee and sugar cane plantations, putting pressure on the government of the newly emerging state to reformulate land policy.

It should also be emphasised that slavery was already in crisis at this time, as a result of the British efforts to expand consumer markets in their areas of influence, the resistance of slaves to captivity and the proliferation of runaway slaves, even in the far corners of the country. In addition, the ruling classes of the economy, specifically the coffee and sugar industries, needed to solve two major problems: legalising land ownership and obtaining labour. The ban on the slave trade was a reality and the masters feared that there would soon be a shortage of labour.

The policy of encouraging the immigration of settlers emerged as a horizon in this process of replacing slave labourers with free men. The question of how these new settlers would be incorporated into Brazilian society remained to be discussed. Beforehand, it was known that coffee producers were not interested in competing with new potential producers. And it was necessary to leave

It was clear that the settlers would come to Brazil to serve the needs of coffee production. Thus, the only way to keep settlers away from land ownership in the short term was to sell it at high prices, as happened with the enactment of the Land Law of 1850.

It is impossible to think about the Land Law of 1850 without analysing the general context of the social and political changes that took place in the first half of the 19th century. On the world stage, European countries such as France and England had undergone a major process of modernisation, both politically and economically. In the previous fifty years, they had been the world's great powers and were living in the throes of capitalist society. In this context, capitalist development had a direct impact on the process of re-evaluating land policy in different parts of the world, such as Brazil and the United States.

In the 19th century, land was incorporated into the commercial economy, changing the relationship with this asset. In this new perspective, land was to become a valuable commodity, capable of generating income both because of its specific character and because of its capacity to produce other goods. The aim was to give land a more commercial character, and not just one of social status, as was typical in the mills of colonial Brazil.

> [...] Its [the Land Law's] main characteristic is that, for the first time, it establishes private ownership of land in Brazil. In other words, the law provides a legal basis for the transformation of land - which is a natural asset and therefore has no value from the point of view of political economy - into a commodity, an object of business, and therefore, from then on, a price. The law then standardised private land ownership (STEDILE, 2005, p. 23).

In Brazil, specifically in the 19th century, coffee replaced sugar as the main product of the agro-export economy. With this replacement, a new class was formed in Brazilian society, an entrepreneurial class that would play a fundamental role in the country's subsequent development (Furtado, 2007).

In this context, it should be noted that after the capital of the Empire was transferred to Rio de Janeiro, it became the country's main consumer market:

> [...] the consumption habits of its inhabitants had changed substantially since the arrival of the Portuguese court. Supplying this market became the main economic activity of the rural population centres that were located in the south of the province of Minas Gerais as a result of the expansion of mining. In this part of the country, the trade in goods and animals for transporting them was the basis of an economic activity of some importance, and gave rise to the formation of a group of local commercial entrepreneurs. Many of these men, who had accumulated some capital in the trade and transport of goods and coffee, became interested in coffee production and became the vanguard of the coffee boom (FURTADO, 2007, p. 170-171).

The Law of 1850 established purchase as the only way of accessing the property.

land and defined which lands were vacant. This law revalidated the sesmarias granted up until 1822, ratified occupations and directly legitimised acquisitions by purchase of land that had hitherto simply been owned, whether possessions or sesmarias. A condition for the revalidation of both possessions and sesmarias was the existence of effective cultivation or the beginnings of effective cultivation on the land sought.

> In this way, the law sanctioned, under certain conditions, all the forms of land acquisition that had existed until then: by government concession (sesmarias), by occupation (posses) and by purchase. All other land, with the exception of that used for public purposes, was considered vacant (SALLUM JR., 1982, p. 15).

The Land Law of 1850 was the culmination of an entire land policy discussed and drawn up during the first fifty years of the 19th century. The beginning of the restructuring of the land code in Brazil began to be thought out together with the policy of integrating the different provinces into a whole that the Brazilian state was endeavouring to combine in an attempt to create the Brazilian nation.

This law was drawn up to play a fundamental role in the transition from slave labour to free labour and, at the same time, to allow the imperial state to regain control over the vacant lands that, since the end of the sesmarias system, had been freely and disorderly transferred to private ownership.

According to Article Three of Law 601 of 18 September 1850, vacant land consisted of:

> §Paragraph 1 - Those applied to any national, provincial or municipal public use;
> §Paragraph 2 - Those that are not in the domain of a private individual by any legitimate title, nor have been granted by sesmarias or other concessions from the General or Provincial Government;
> §Paragraph 3 - Those that are not given by sesmarias, or other government concessions that, despite being subject to commissions, are revalidated by this Law.
> § Paragraph 4 - Those that are not occupied by possessions, which despite not being based on a legal title, are legitimised by this Law (BRASIL, 1850).

It is worth highlighting the first article of this law, which prohibited the acquisition of vacant land by any means other than purchase, as well as the fourteenth article, which made the imperial government the "seller" of vacant land, setting a minimum price for it, which at the time was higher than that of private land. Article 18 authorised the government to bring in immigrants to work in agricultural establishments, in services run by the public administration or to form colonies. Finally, the nineteenth article stipulated that the proceeds from the sale of land should be used to measure vacant land and to import settlers (SALLUM JR., 1982, p. 15).

As a result, this law transformed vacant land into a state monopoly, controlled by the class of large landowners.

The central aim of the imperial government's policy was to demarcate wastelands for sale and use in colonisation projects, as a way of financing the arrival of immigrant workers for the farms, thus remedying the possible shortage of labour with the end of the slave trade. But for this to happen, it would be necessary to put an end to territorial appropriation through possession.

However, as Silva (1996, p. 335) states, "the squatters' refusal to demarcate their land and legalise their titles prevented the implementation of the imperial project of colonisation with small property". We can affirm that the Crown's objective was never linked to combating large estates and slavery, but rather to regaining control, which it had long since lost, in the form of territorial appropriation.

The strength of the landowning class, which organised the Brazilian territory into large estates, is thus evident, firstly by adapting to the objectives of colonial exploitation and, a posteriori, by the economic status brought by land ownership. The pact that began in the colonial period remained latent and stronger after the formation of the national state, since the constitution of the landowning class took place at the same time as, and was intertwined with, the process of consolidating it (SILVA, 2008).

Without demarcating the wastelands, the problem of a lack of labour was solved through subsidised immigration, in which both the state and the farmers competed, helping the immigrants with housing and a small plot of land so that they could produce for subsistence. However, everything that was subsidised to the immigrant had to be paid for later, either through services or payment in kind (money or produce).

Therefore, the creation of this law reaffirmed land concentration and aimed to create obstacles to small rural property. If these obstacles did not exist, the immigrant would become the owner and it would be practically impossible to obtain workers without the coercion of slavery. Therefore, as Martins (1999) states, in a country with a free labour regime, land had to be captive in order to serve as an instrument of domination.

> The Land Law of 1850 already had an ambiguously conservative character, which shows that, deep down, the big landowners were gradually building up and strengthening their power [...]. Far from aiming to liberalise access to land, the Land Law aimed to do just the opposite: to block workers' access to property, so that they would compulsorily become the labour force of the large farms (MARTINS, 1999, p. 76).

The Land Law was closely linked to the process of consolidating the national state and the formation of classes, insofar as it sought to order a tumultuous situation that existed in terms of property titles in the period between the end of the concessions and the implementation of the 1850 Law, which took place in 1854.

Based on Silva's (1996) observations, we can say that the law established a new relationship between landowners and the state, which evolved during the second half of the 19th century. The composition of this new relationship between landowners and the state was fundamental, since landlords had always played a strong role in the social and political organisation of the Imperial State.

Analysing the drafting and implementation of the Land Law reveals that the landowning class was always resisting, pressurising and adapting the legislation to their most pressing interests.

So:

The law was drawn up as part of an overall project for society [...], but its application to society was the result of a process in which the different social strata concerned came into conflict and found the means to accommodate the legal system to their interests (SILVA, 1996, p. 343).

The Republican Constitution of 1891, Article 64, transferred vacant land to the newly created states, which meant concentrating land control in the hands of the state oligarchies.

Since then, although the Law of 1850 has remained the basic law for land management, each state has developed its own land and labour policy, initiating the transfer of land to large farmers and colonising companies, generating intense real estate speculation. One of the first examples of land management, already the responsibility of the states, was Law No. 323 of 22 June 1895 in São Paulo:

[...] It provided for vacant lands, their measurement, demarcation and acquisition. On the legitimisation or revalidation of possessions and concessions, discrimination of public and private domain, as well as other provisions (SALLUM JR, 1982 p. 17).

The transfer of wastelands to private ownership was intrinsically linked to the typical phenomenon of the First Republic (1889-1930), colonelism.

Silva (1996, p. 336) points out that:

> By controlling municipal life through means that ranged from paternalism to violence, the colonels "loyal" to the oligarchies that dominated state politics played a central role in the way in which wastelands were incorporated into private patrimony.

Coronelismo was a system of political power characterised by enormous power concentrated in the hands of a powerful local, usually a large landowner, but also, not infrequently, colonels could be merchants who traded in agricultural products and bought the produce of landowners. Thus, according to Martins (1999, p. 28), "the whole system was [...] based on mechanisms of political intermediation with a patrimonial foundation".

Institutional coronelismo emerged with the formation of the National Guard, created in 1831 as a result of the deposition of Dom Pedro I in April of that year. As the National Guard was inspired by the French institution, the "bourgeois guard", forged by the events of 1789, was a civilian militia that represented the armed power of the landowners who began to patrol the streets and roads in place of the traditional forces, overthrown by the revolutionaries.

To be a member of the National Guard, you had to be well-off and have the resources to pay for the uniform and the necessary weapons.

> [...] in the Republic [...] the National Guard ended up playing an essential role. Its members were ranked according to a military hierarchy, being called "colonels", "majors", "captains", etc. The municipal or regional political leaders ended up being known as "colonels" and the political phenomenon they marked (sic) with their presence became known as coronelismo (MARTINS, 1986, p. 46, emphasis added).

In the political-administrative sphere, coronelismo was characterised by the rigid control of the political bosses over the votes of the electorate, constituting the "electoral corrals[2] " and producing the so-called "cabresto vote", in other words, "the voter and his vote were under the tutelage of the colonels, who disposed of them as their own" (MARTINS, 1986, p. 46).

---

[2] The voter exchanged their vote for a favour, which could be a material good (shoes, clothes, hats) or some kind of gift (medical care, medicine, burial money, a doctor's appointment, school enrolment, a scholarship). This obedient placidity of those who had the right to vote made them part of the electoral corral. By behaving like puppets in the elections, it was inevitable that they would be seen as inferior people, incapable of reacting to the despotism of the colonel.

So:

> As a representative of local bossism, the colonel exerted his influence - paternal if possible, coercive if necessary - over the poorer sections of the population, who often lived as servants on the lands of the powerful locals. In this way, the colonel provided a service to state politicians during election periods, recruiting his "halter" voters and expecting favours from state politics for his municipality and his person in return. (SILVA, 2008, p. 279)

So, over time, the colonel automatically came to be seen by the people as a powerful man on whom everyone else depended.

With the advent of the Republic, there was little expansion of citizenship rights, implicit in the abolition of slavery. Paradoxically, it prohibited the electoral rights of freedmen, while eliminating the restrictive precept of minimum income for political participation and excluding illiterates from the right to vote.

On the subject, Martins (1986, p. 45) states that "this did indeed broaden electoral participation, but not to the same extent as the formal equality[3] resulting from the generalisation of free labour".

Under the government of Campos Sales (1898 - 1902), the "governors' policy" began, based on supporting the presidency of the Republic and, reciprocally, the governors through a system of exchanges of political favours. The governors operated within the same system of exchanges with the colonels. This system involved appointing municipal officials, police and judicial authorities on the recommendation of the colonels, as well as facilitating land concessions and public works. Thus, Martins (1986, p. 46) states that, "with this, each political leader in tune with the state government became a true municipal ruler, setting and ruling over everything and everyone".

The electorate of the colonels was made up of their clients, usually small traders and farmers. The vote was treated as a commodity. In exchange for the voter's loyalty, the colonel could offer anything from shoes to pieces of land. In this way, according to Martins (1986, p. 47), "coronelismo enshrined an effective system of

---

[3] Regarding formal equality, a basic precept for free labour, Rodrigues (2006, p.66) points out that although production is increasingly social, it is mediated by acts of exchange, taking the form of relations between private individuals. In this way, while exploitation and despotism are concealed in the sphere of production, the representation of a society of free individuals is exacerbated in the sphere of circulation. This sphere where the buying and selling of labour power takes place appears as a veritable paradise "of the natural rights of man", where freedom, equality and property reign; where the buyer and seller of labour power "contract as free people, legally equal"; where the "only power that brings them together and leads them to a relationship is self-profit, private advantage, their private interests". Everyone is equal in the sense that they are responsible for their own actions and, as long as they can buy or have something to sell, they participate in the market on equal legal terms.

political exclusion of all dissidents who could not move a clientele to negotiate political positions with them". In order to assert their regional power, the colonels had at their disposal a large number of jagunços, labourers and servants.

Like this:

> Coronelismo entangled land issues, honour issues, family issues and political issues in a complicated web. The old family wars, which had been going on since the colony, were amplified, now complicated by party-political issues. The colonel's power was therefore not his own, but that of the government which he supported electorally and which supported him politically. Therefore, his power basically depended on his ability to barter. This means that patronage relations were governed by a peculiar system of equality, which was tied equality, constituted by the exchange of favours for votes - a market equality, which only takes place between owners of goods (MARTINS, 1986, p. 48-49).

As a result, the power pact between the state and landowners at this time led the political oligarchies in Brazil to place, according to Martins (1999, p. 20), "[...] the institutions of modern political domination at their service, submitting the entire state apparatus to their control."

Therefore, it is indisputable that the state's performance, in general, was geared towards serving the interests of the landowning class. This is even more evident in the café con leche policy, which began in 1898 in the Old Republic, when there was alternation in government between the oligarchies of São Paulo and Minas Gerais, lasting until the 1930 Revolution.

In this context, we cannot fail to highlight two major events that changed the direction of Brazilian politics and economics: the 1929 crisis and Brazilian industrialisation.

According to Furtado (2007, p. 252), the first coffee overproduction crisis began in the early years of the 20th century:

> [...] Brazilian entrepreneurs soon realised that they were in a privileged position among producers of primary goods to defend themselves against falling prices. All they needed were the financial resources to keep part of their production off the market, in other words, to artificially contract supply. The stocks thus formed would be mobilised when the market showed more resistance, i.e. when incomes were at high levels in the importing countries, or they would be used to cover shortfalls in years of bad harvests.

Even with the possibility of a partial stockpiling mechanism for coffee production, overproduction of the product did not cease and stocks increased year on year, weighing on prices, which caused a permanent loss of income for producers and the country. Thus, in 1906, a convention was held in the city of Taubaté, where the foundations of the coffee "valorisation" policy were defined. In essence, this policy consisted of the following:

> a) In order to re-establish the balance between supply and demand for coffee, the government would intervene in the market to buy up the surplus;
> b) These purchases would be financed with foreign loans; c) the servicing of these loans would be covered by a new tax levied in gold on each bag of coffee exported;
> d) in order to solve the problem in the longer term, the governments of the producing states should discourage the expansion of plantations (FURTADO, 2007, p. 254).

The controversy surrounding this policy pointed to the transformations that were taking place in the country's political and social structure, since the power of coffee producers at regional level had been strengthened by republican decentralisation and by the implementation of the banking reform, which made it possible to expand credit, benefiting the coffee-growing class in two ways: firstly, to finance the opening up of new land, and by raising the price of coffee in national currency due to the depreciation of the exchange rate.

In addition, there was a growth in both the number and complexity of the groups that exerted pressure on the central government. Also noteworthy was the growing importance of the urban middle class, including the civil and military bureaucracy, which was directly affected by the depreciation of the exchange rate.

> The important international financial group gathered around Rothschild closely followed the economic and financial policy of the Brazilian government, particularly after the consolidation loan of 1898. 'Rothschild feared the policy of "valorisation" of coffee, because if the Brazilian government went bankrupt, this could have repercussions on the payment of the foreign debt'. Finally, the importing merchants and industrialists, whose interests for different reasons were opposed to those of the coffee growers, found in the republican regime an opportunity to increase their political power (FURTADO, 2007, p. 254-255, emphasis added).

The coffee economy's defence mechanism worked relatively efficiently until 1929. However, the global crisis that ensued left Brazil extremely vulnerable. The reason for this was the large coffee production due,

mainly due to the artificial stimuli received and, on the other hand, the fall in imports. The ample supply of coffee was not absorbed by the market, as demand grew slowly.

There was therefore a perfectly characterised situation of structural

> imbalance between supply and demand. We couldn't expect an appreciable
> increase in demand as a result of a rise in disposable income for
> consumption in the importing countries. Nor could we think of increasing
> consumption in these countries by lowering prices. The only way to avoid
> huge losses for the producers and the exporting country was to prevent - by
> withdrawing part of the production from the market - supply from rising
> above the level demand required to maintain a more or less stable per capita
> consumption in the short term. It was perfectly obvious that the stocks that
> were accumulating had no chance of being used economically in the
> foreseeable future. Even if the world economy managed to avoid another
> depression after the great expansion of the 1920s, there was no door
> through which we could foresee the exit of those stocks, since productive
> capacity continued to increase. The situation that had been created was
> therefore absolutely unsustainable (FURTADO, 2007, p. 258).

Even with the precarious situation shown by the author, overproduction did not cease, and retaining the product was impractical, since obtaining credits abroad to finance this retention was impossible, as the international capital market was in a deep depression. The way out chosen by Brazilian coffee growers was to continue harvesting coffee and keep up the pressure on the market, which led to a further fall in prices and a further depreciation of the currency, contributing to aggravating the crisis.

The only way to restore the balance between supply and demand, at a higher price level for coffee and to protect the coffee sector, was to remove part of the harvested coffee from the market in order to destroy it.

> By guaranteeing minimum purchase prices that were remunerative for the
> vast majority of producers, they were actually maintaining the level of
> employment in the export economy and, indirectly, in the producer sectors
> linked to the domestic market. By avoiding a contraction of large portions
> of the monetary income of the export sector, the effects of the
> unemployment multiplier on the other sectors of the economy were
> proportionally reduced (FURTADO, 2007, p. 268).

In this way, the policy of defending the coffee sector was presented as a programme to boost national income. However, it cannot be assumed that this policy was not intended precisely to maintain the flow of income in the country. The displacement of the flow of income from the export sector to industry e,

Consequently, the recovery of the Brazilian economy after the 1929 crisis is explained by trade.

It should also be noted that Brazil already had small industries, mainly textiles and food, which catered for the domestic market and quickly overcame the effects of the great depression of the first half of the 20th century, as well as agricultural production aimed at the domestic market.

Activities linked to the domestic market not only grew, boosted by higher profits, but also received a new impetus by attracting capital that was formed or de-inverted in the export sector (FURTADO, 2007, p. 278).

These activities date back to the end of the 19th century, with the start of significant industrialisation in Brazil, in which the landowning class led the economic changes in the country. Many coffee growers began to invest part of the profits they made from exporting coffee in establishing industries, mainly in São Paulo and Rio de Janeiro; thus, Brazilian industry was born as a subsidiary to the main economic activity of the time.

After the crisis of the capitalist system in 1929, Brazilian industry gained significance due to the shift in the flow of income from the coffee exporting sector to commerce and industry, as well as the arrival in 1930 of the first president to come from the agrarian oligarchy. However, he sympathised with and was tied to the interests of the nascent urban-industrial bourgeoisie, thus assuming a more nationalist and industrialist character.

In 1931, Vargas announced his determination to set up basic industries. With them, the country would be able to reduce its imports, stimulating domestic production of consumer goods.

As a result, a national market was formed for industry, due to the greater fluidity of the territory, with the construction of ports, railways and motorways, facilitating the movement of goods, people and capital.

This period saw the beginning of the transition from an agro-export economy to an urban-industrial economy, which would have its strongest expression under the Vargas government, with the implementation of the import substitution model in an attempt to strengthen national industry.

The context of Getúlio Vargas' accession to state power was marked by the weakening of the agrarian oligarchies, mainly due to the economic crisis of 1929, which paved the way for the ideals of modernisation and development, linked to the industrial-based economy, which had gained strength after the crisis.

In this way, we can understand that the crisis of the oligarchies was a crucial step towards the "revolution" of 1930. With the impact of the 1929 crisis, the then president of São Paulo, Washington Luís, decided to support the candidacy of his fellow countryman, Júlio Prestes. Known as the "Pure Coffee Policy", Júlio Prestes' candidacy broke with the old arrangement of the "Coffee with Milk Policy", in which landowners from Minas Gerais and São Paulo would alternate in the presidential mandate.

Dissatisfied with this measure, a group of dissident oligarchs, mainly from Minas Gerais, Rio Grande do Sul and Paraíba, created an electoral slate against the candidacy of Júlio Prestes. Known as the Liberal Alliance, the slate led by the then governor of Rio Grande do Sul, Getúlio Dorneles Vargas, promised a set of reformist measures. Its programme presented some progressive advances, such as the establishment of labour legislation, an eight-hour working day, a secret ballot, women's suffrage, support for the urban classes and the development of national industry. However, in the 1930 elections, the Liberal Alliance lost to the Republican candidate Júlio Prestes. Using the assassination of alliance member João Pessoa by a Washington Luís sympathiser as a pretext, Getúlio Vargas and his supporters organised a coup that removed Washington Luís from power in October 1930, putting an end to the so-called Old Republic.

At the start of his government, with the centralisation of power, Vargas began the fight against regionalism. The administration of the country should be a single one and not as it had been in the Old Republic. "[...] the presidents of the Republic were generally supported in central power to the extent that they recognised the independence and local and regional power of the political bosses, the 'colonels' of politics" (MARTINS, 1984, p. 21).

Even with the military in place of the old regional political bosses, some invested with great personal power, the government was unable to decimate political patronage,

> [...] the oligarchies maintained their relations of patronage and, above all, their traditional domination over their clientele, i.e. the people. But they began to obey the new masters of power, the military and bureaucrats of the centralised state. In the end, even where the Revolution renewed political leadership, it resorted to the same system of compromises with local factions on which coronelismo had always been based (MARTINS, 1999, p. 31-32).

In order to control the overproduction of coffee and the crisis in Brazil, Vargas incinerated the stocks of the product. Despite the global crisis, industrial development accelerated. This development was due to the reduction in imports and the supply of capital, which led many coffee growers to switch from traditional farming, which was in crisis, to industry. However, it was the participation of the state, through protectionist tariffs, as well as the policy of defending the coffee sector and investments, that had the greatest influence on industrial growth. Unlike in the Old Republic, plans began to emerge for the creation of basic industries in Brazil. These plans were realised with the inauguration of the Volta Redonda steel plant in 1946.

This is the context of the changes expressed by the power pact redefined from the 1930s onwards. The establishment of the industrial bases of the Brazilian economy was accompanied by changes in the way of thinking, values and beliefs of Brazilian society, such as the measures to create a public education system, officially controlled by the government, compulsory religious education in public schools and the creation of the Ministry of Education and Health. With regard to higher education, the government sought to establish the foundations of the university system, investing in the areas of teaching and research. As well as educational development, there was a real cultural revolution compared to the Old Republic. Modernism, so criticised before 1930, became the main artistic movement after Vargas' coup.

The process of urbanisation accelerated with the advance of industrialisation, which led to a huge growth in the working class. Vargas, with a government policy aimed at urban workers, tried to attract the support of this class, which was fundamental to the economy. The creation of the Ministry of Labour, Industry and Commerce in 1930 resulted in a series of labour laws. Some of these were aimed at extending workers' rights and guarantees: holiday laws, regulation of women's and children's work. Various transformations that would, in theory, lead to modernisation took place in the social, economic and political spheres. However, these transformations were more restricted to the city. This is understandable, because even with the rise of the bourgeoisie as a class and as the leader of the state, the interests of landowners had to be guaranteed. Martins (1999, p. 72) states that:

> It is significant that the same Getúlio Vargas who proposed and made possible the Consolidation of Labour Laws in 1942, to regulate labour issues in factories and cities, did not extend legal rights to rural workers that would give a contractual form to labour relations that were still strongly based on criteria of personal dependence and true servitude. As a result, Vargas was unwilling or unable to confront the big landowners and their allies. It was during his government that the foundations were laid for a tacit political pact, still in force today, with modifications, in which the landowners do not run the government, but are not opposed by it (emphasis added).

If until the beginning of the 20th century, the agrarian oligarchies had complete political and economic power, the scenario changed, as already mentioned, from 1929 onwards with the drastic fall in the price of coffee on the world market, and with it the loss of prominence of the agrarian oligarchies in the state apparatus. From the 1930s onwards, Brazil redefined the axis of economic accumulation, shifting from coffee monoculture to industrial production. In this way, the country ceased to be essentially an agrarian exporter and became an urban-industrial country.

Until the end of the Old Republic, the agro-export pact was hegemonic, in which the entire state apparatus was geared towards meeting the needs of the coffee exporting sector and, therefore, the landowning class. With the advent of significant industrialisation and the entry into the state administration of a president closer to the interests of the nascent urban-industrial bourgeoisie, a new pact became hegemonic, the bourgeois pact, based on industry, the city and the ideology of progress and development. However, the Brazilian bourgeoisie came from the agrarian oligarchy, so a rupture of the previous pact and a clash of classes was unthinkable. That's because:

> [...] unlike in other societies, the elite did not have an antagonistic class that was sufficiently strong and aware of its interests and oppositions, such as an industrial bourgeoisie or simply a modern bourgeoisie opposed to the interests of the latifundia, which could carry out social reforms that did not affect fundamental political and ideological choices. (MARTINS, 1999, p. 74)

Paulino (2009, p. 81) goes on to say:

> [...] in the classical model, the bourgeoisie established itself as a counter-hegemonic force to the power structures remaining from the feudal order, whereas in Brazil, which was formed under the aegis of commercial capitalism, a significant part of the wealth under the control of the agrarianists involved in the agro-export economy was directed towards urban-industrial activities, at first precisely as a strategy to increase earnings in agricultural activity. With this, some of them personified two class situations: landowners and, at the same time, urban-industrial entrepreneurs, bourgeois in short.

Since there was no effective break when the urban-industrial bourgeoisie came to power, as in the classic model, the agro-export pact ceased to be hegemonic, but the interests of the landowning class were secured by the bourgeois-owners. For this reason, the land structure in this country has never been significantly altered.

> [the 1930s] was a time of particular demarcation of roles in a society that was, in theory, modernising [...] a completely different positioning in the game of hegemonic forces was made explicit than that seen in the countries that were the protagonists of the bourgeois revolution, which emerged precisely from the victorious confrontation between the bourgeoisie and the landowners in the early days of the Industrial Revolution (PAULINO, 2009a, p. 77).

According to Paulino and Almeida (2010, p. 82), in the classic model of capitalist development in the central countries, the bourgeoisie only became the dominant class when it managed to undermine the power of landowners, who were seen as an obstacle to the accumulation of industrial capitalism. "This is because the original accumulation formula presupposes the appropriation of surplus value by reducing the cost of reproducing the labour force, with food being a not insignificant factor in the composition of these costs."

The clash of classes between the bourgeoisie and landowners took place in the early days of the Industrial Revolution, and the culmination of the rupture was the end of the Corn Laws in England in 1846.

According to Paulino and Almeida:

> [...] enacted [the Corn Laws] in the midst of disputes between industrialists, cornered by the pressure to increase wages due to food prices, and landowners, busy guaranteeing a monopoly on supplying the domestic market and, with this, the potential appropriation of land rent (PAULINO; ALMEIDA, 2010, p. 82).

The distinction between profit and income should be emphasised at this point. Profit stems from the appropriation of the value created by surplus labour (labour not converted into wages) and necessarily presupposes capitalist investment as the direct means of its extraction. Rent, on the other hand, originates from the enclosure of land, which is transformed into a private means of production without requiring any investment (PAULINO; ALMEIDA, 2010, p. 82, 83).

Like this,

> [...] in capitalism, landowners are vested with the right to levy a tax on everyone who needs to live, feed, clothe and live, because all of this requires land. Unlike profit, which originates directly from the labour relationship, rent is extracted indirectly, because in the end it is the surplus labour that will remunerate its owners. It is a social tax, therefore, and one that grows as all kinds of demands increase, which can be satisfied through the mediation of this material base that is irreplaceable to every form of life (PAULINO; ALMEIDA, 2010, p. 83).

While the monopoly of land ownership in the central countries of capitalism represented an obstacle to its complete development, the democratisation of access to land was necessary to boost the capitalist economy; in Brazil, this obstacle was overcome by the land-capital alliance, the personification in one person of the large landowner and the capitalist, so there were no clashes between the nascent industrial bourgeoisie and the rural oligarchies, as in the European case, a break with the landowners, transferring the centrality of accumulation to the circuit of capitalist production, while at the same time maintaining control over land ownership, preventing the looting of income from threatening average profit rates (PAULINO; ALMEIDA,

2010, p. 83).

Thus,

> Unlike the classic model of the relationship between land and capital, in which land (and land rent, i.e. the price of land) is recognised as an obstacle to the circulation and reproduction of capital, in the Brazilian model the obstacle to the capitalist reproduction of capital in agriculture has not been removed by land reform, but by tax incentives. [...] The Brazilian model inverted the classic model. In this sense, it politically reinforced the irrationality of land ownership in capitalist development, consequently strengthening the oligarchic system based on it (MARTINS, 1994 apud PAULINO; ALMEIDA, 2010, p. 85).

Thus, in the classic model of capitalist development,

> The distribution of land and the definition of limits to land concentration were fundamental in weakening oligarchic power and, at the same time, increasing the supply of food, which provided the capitalists with the largest share of the value derived from the surplus labour extracted from the workers. This is what has ensured the supply of food at prices that don't compromise the constitution of a solid demand for other durable and non-durable consumer goods; in short, the path by which an internal market was consolidated that the capitalists couldn't do without, not even in these times of the globalisation of capital (PAULINO; ALMEIDA, 2010, p. 83).

Deepening the discussion on this issue, the authors point out that:

This alliance results in a shift in the dynamising power of the economy, of production, towards private ownership of land, which diverges from the classic model of capitalism, in which the mechanisms of accumulation are based precisely on the negation of income as the primary element in the process of accumulation (PAULINO; ALMEIDA, 2010, p. 82).

This is what happened,

> [...] the great social and economic changes in contemporary Brazil are not related to the emergence of new social and political protagonists, bearers of a new and radical political and economic project. The same elites who were responsible for the level of backwardness in which they found themselves in a previous historical situation, were the protagonists of the social transformations. (MARTINS, 1999, p.58)

With the rise of the bourgeoisie in control of the Brazilian state, it was hoped that agrarian reform would be carried out in order to develop capitalism and completely modernise the country. However, the land-capital alliance had already laid its foundations,

> Between the old elites and the new elites, a kind of political compromise was established, whereby the industrialists and big businessmen became

<blockquote>
the oligarchies' major political clients, to whom they delegated their responsibilities of command and direction, reproducing the same political mechanisms that victimised all the people and prevented the effective development of democracy among us. (MARTINS, 1997, p. 20)
</blockquote>

On this subject, Stedile (2005, p. 28-29) points out that:

<blockquote>
The political elites - the industrial bourgeoisie, now in power - made an alliance with the rural oligarchy [...] to maintain it as a social class, for two fundamental reasons: firstly, because the Brazilian industrial bourgeoisie originated from the rural oligarchy, from the accumulation of coffee and sugar exports, unlike the historical processes that took place in the formation of capitalism in Europe and the United States. The second reason: as the industrial model was dependent, it needed to import machines, and even labourers, from Europe and the United States. And the import of these machines was only possible because of the continuity of agricultural exports, which generated foreign currency to pay for them, closing the cycle of the logic of necessity of dependent capitalism (STEDILE, 2005, p. 28-29).
</blockquote>

The landowning class had its power affected and no longer directly controlled the country's politics and economy, but the permanence of this class, albeit indirectly but emblematically, was reflected in the maintenance of the land structure and the failure to implement the social and wage policies developed for the urban-industrial sector in the rural sector (MARTINS, 1999).

With the end of the Vargas government in 1945, the country was governed by the PSD (Social Democratic Party), which clearly represented oligarchic interests and decided the course of power throughout the democratic period between 1946 and 1964.

The 1946 Constitution, which was supposed to be more democratic, reinforced the power pact between large landowners and the state, since land expropriated for social purposes, including land reform, had to be paid for in advance and in cash to the owner, thus making land reform economically unviable.

In Juscelino Kubitschek's government, the innovations brought about were like a "doubling of the state machine", as Martins (1999, p. 59) points out, with the creation of executive groups (such as the Executive Group of the Automobile Industry - GEIA) and development agencies, such as the Superintendence for the Development of the Northeast (SUDENE). In the latter case, the interests of clientelist and oligarchic groups in the Northeast were combined with the modernising interests of businessmen in the Southeast.

Thus,

<blockquote>
The Kubitschek government did not abolish archaic political patronage
</blockquote>

During this period, there was considerable industrial progress, mainly in the basic industry sectors and in the production of durable and non-durable consumer goods. The government sought to attract foreign capital to invest in the country, obtaining loans and encouraging international companies to set up in Brazil, such as Ford, Volkswagen, Willys Overland and General Motors (GM).

The international situation favoured these investments, as developed countries such as the United States had a good reserve of capital available.

However, the economic progress of this era presented a number of problems: the concentration of wealth in the Southeast, the increase in foreign debt and the growing devaluation of the Brazilian currency.

Juscelino Kubitschek's Target Plan was the responsibility of the Development Council, based on the economic planning effort developed by the Brazil - United States Joint Commission and later by the BNDES (National Bank for Economic and Social Development) - ECLAC (Economic Commission for Latin America and the Caribbean) Joint Group. The Plan identified the sectors that, if properly stimulated, would be capable of growth and remove possible bottlenecks in terms of infrastructure (energy and transport).

There were thirty specific goals, divided into five sectors: energy, transport, basic industry, food and education. In addition to these, there was an autonomous goal: the construction of Brasilia. The plan's main forms of financing were monetary expansion to finance public spending and credit for private investment, making monetary and fiscal policies passive and subordinate to the structural reforms of the economy.

As far as agriculture was concerned, the aim was to expand production and improve productivity, but what happened was a crisis in the bean and wheat plantations.

In the Plano de Metas, not enough importance was given to agriculture and no importance was given to the issue of land reform, even though there were many studies on the subject, because the emphasis of the plan was on industrial development. It's worth highlighting the huge failure of the policy announced for wheat, as the country didn't have the phytogenetic research technology to allow adapted seeds to emerge.

As far as agriculture is concerned, we would highlight the expansion of the agricultural frontier, made possible by the widening of the road network and important migratory flows (rural - rural), such as from the north-west of the state of Rio Grande do Sul to the west of Santa Catarina; the south-east and west of Paraná and the south of Mato Grosso. We can see an increase in public and subsidised agricultural credit for the production of coffee, rice, wheat, cotton and peanuts.

Although, in general terms, the policy of overvaluing the exchange rate penalised agriculture, in the 1950s the branches of agricultural production that were able to modernise technologically were favoured by the overvalued exchange rates, as they were able to import inputs, especially fertilisers, tractors, trucks and oil derivatives at subsidised costs.

The goal of building railways also failed, although that of roads far exceeded what had been planned.

Another aspect that should be mentioned in the context of the Target Plan was the Brazilian government's break with the International Monetary Fund (IMF) in 1959, due to the failure to adopt a stabilisation policy in line with the Fund's requirements, which wanted to contain government spending (including halting the construction of Brasilia) in order to stop rising inflation.

With the arrival of transnational capital, an important union was formed in the country between three pillars that stimulated the economy: national private capital, state-owned companies and multinational companies.

In the late 1950s, the countryside began to receive technological innovations from industry, stimulated by the federal government through selective, subsidised financing from the Bank of Brazil for large landowners. Land ownership conditioned access to financing and production tools (fertilisers, tractors, trucks and inputs in general) which became, in the mindset of the government and large landowners, the real means of production. This was a great move, because since land is the real means of production in agriculture, transferring the importance of production to production tools shifted the cause of the country's economic backwardness to technological deficiencies and not to the concentration of land, removing land from the focus of discussion. This resulted in reinforcing the role of land as a store of value and a source of economic power.

The aim of this government action was to eradicate the old coffee plantations and replace them with other crops, so that the machinery produced by the industries could be used. This practice became more evident in the 1960s, when the combined interests of the rural oligarchies and the industrial bourgeoisie culminated in the conservative modernisation of some large estates.

A package of incentives and the mobilisation of huge resources promoted the gradual replacement of human labour by machines and implements. This reinforced land concentration and the expropriation and expulsion of peasants who had precarious possession of the land.

At the same time, in the mid-1950s, contesting the great social inequality, land concentration and the process of expropriation that existed in Brazil, the Peasant Leagues were formed and rural unionisation began:

> [...] it was with the Peasant Leagues, in the 1940s to 1960s, that the struggle for agrarian reform in Brazil took on a national dimension. Often born as a charitable society for the dead, the Leagues went on to organise the struggle of the landowners, residents, tenant farmers, small landowners and salaried rural workers of the Zona da Mata against the latifundia, especially in the north-east of Brazil. (OLIVEIRA, 2007, p. 105)

The Peasant Leagues, according to Martins (1986, p. 76-78), emerged in 1955 in the Galiléia plantation in Pernambuco, from an association of landowners called the Agricultural and Livestock Society of Pernambuco Planters.

> [...] The leagues spread rapidly throughout the Northeast, initially with the support of the Communist Party of Brazil and with severe opposition from the Catholic Church. They emerged and spread mainly among the landowners of old mills that were beginning to be taken over by their absentee owners, due to the rise in the value of sugar and the expansion of sugarcane plantations. Since the 1940s, the "foreiros" had been evicted from the land or else, as we have seen, reduced to "moradores de condição", a step towards becoming non-resident salaried workers (MARTINS, 1986, p. 76).

Martins (1986, p.77-78) also states that the emergence of the Peasant Leagues took place in a context that was not only broader than the expulsion of the landowners and the reduction or extinction of the plantations of the mill dwellers, but also in a context of regional political crisis.

According to the author:

> This crisis was characterised by an awareness of the underdevelopment of the Northeast and, in particular, a definite move by the regional bourgeoisie to obtain from the federal government not just a paternalistic policy of

emergency aid in times of severe drought, but an effective economic development policy. In other words, a policy of industrialising the Northeast. The problem of the misery of the peasants and their exodus to the south was explained as the result of under-utilised latifundia, which prevented those who needed the land from occupying it. A regional development policy based on industrialisation was supposed to halt and reverse the vicious circle of poverty caused by monoculture and landowning agriculture. This is how the Superintendence for the Development of the Northeast came about, and how political alliances emerged involving such opposing extremes as the Communist Party and the National Democratic Union, the party par excellence of the bourgeoisie. In Pernambuco, this 'centre-left' alliance made it possible for Cid Sampaio, a mill owner, to win the Recife City Council and later the state government. Despite the opposition of the plantation lords, now reduced to the status of mere cane suppliers to the powerful sugar mills, the peasant leagues and, soon afterwards, a strong rural unionisation movement, took hold in the region, guaranteed at first by the political weakening of these former colonels. There were two distinct groups of workers to be mobilised and organised. On the one hand, the landowners of the sugar mill lands, peasants on the verge of expulsion. On the other hand, the residents of the mills, workers who were in the process of definitively becoming wage earners, losing their peasant characteristics, as well as those who had already been effectively reduced to the condition of wage earners, expelled from their fields to the ends of the streets, the villages near the mills (MARTINS, 1986, p.77-78).

The tensions and demands entered the 1960s when the Brazilian peasantry began to become more organised, although not at a national level. Movements to democratise access to land gained momentum under João Goulart's government, as this president with his progressive ideas believed that land reform was essential for the complete modernisation of the country. To this end, Law No. 4.132[4] was enacted on 10 September 1962. This law defined the cases of expropriation in the social interest and, "[...] from a legal point of view, it was a significant step towards passing the first law on land reform in Brazil" (OLIVEIRA, 2007, p. 112).

Also noteworthy during Goulart's government was the creation of SUPRA (Superintendence of Agrarian Policy) in 1962 and the approval of Law No. 4,214[5] , of 2 March 1963, known as the Rural Workers' Statute, which allowed for the implementation of rural trade unionism, as set out in Article 114.

---

[4] Art. 1 Expropriation in the social interest shall be decreed in order to promote the fair distribution of property or to condition its use to social welfare, in accordance with Art. 147 of the Federal Constitution.
    Art. 2 is considered to be of social interest:
I - the use of all unproductive or exploited property that does not correspond to the housing, work and consumption needs of the population centres that it must or can supply through its economic use;
§ Paragraph 1 of item I of this article shall only apply in the case of goods taken out of production or in the case of rural properties whose production, due to inefficient exploitation, is lower than the average for the region, taking into account the natural conditions of their soil and their situation in relation to the markets. (OLIVEIRA, 2007, p. 113).
[5] Art. 114: It is lawful for all those who, as employees or employers, carry out rural activities or professions to join a trade union for the purposes of studying, defending and coordinating their economic or professional interests (OLIVEIRA, 2007, p. 117).

With regard to SUPRA, as noted by Oliveira (2007, p. 114), we highlight Article 2 of Delegated Law No. 11 of 11 October 1962, which provides for its creation and powers.

> Art. 2 SUPRA is responsible for collaborating in the formulation of the country's agrarian policy, planning, promoting, executing and enforcing, under the terms of the legislation in force and that which may be issued, agrarian reform and, on a supplementary basis, the complementary measures of technical, financial, educational and sanitary assistance, as well as others of an administrative nature that may be conferred on it in its regulations and subsequent legislation. In order to promote the fair distribution of property and condition its use to social welfare, SUPRA is delegated special powers of expropriation, in accordance with the legislation in force.

Still under João Goulart's government, it is necessary to recall some passages from the public presentation of the Agrarian Reform Project, in the speech given on 13 March 1964 at the Central Station in Brazil:

> Those who imagine that the government will be able to stifle the voice of the people or stifle their demands are wasting their time. Those who fear that the government will take subversive action in defence of political or personal interests are also wasting their time, as are those who expect the government to take repressive action against the people, their rights or their demands (JOÃO GOULART apud STÉDILE, 2005, p. 100).

> The people want [...] land ownership to be accessible to all; for everyone to be able to participate in the political life of the country by voting and being able to vote; for the intervention of economic power in the electoral process to be prevented, and for the representation of all political groups to be guaranteed, without any discrimination, ideological or religious (STÉDILE, 2005, p. 101).

Goulart also said that this Superintendence could not be considered the agrarian reform that peasants were fighting for and hoping for, but he believed it was a first step. He also questioned how payment for expropriations would be made, saying that the Agrarian Policy Superintendence would not be enough to effectively carry out agrarian reform.

> [...] the Supra decree [...]. It's still not the agrarian reform we've been fighting for. It's still not the overhaul of our impoverished rural landscape. It's not yet the letter of release for the abandoned peasant. But it is the first step: a door that opens to the definitive solution of Brazil's agrarian problem (STÉDILE, 2005, p. 103 - emphasis added).

> Agrarian reform with prior payment for unproductive latifundia, in cash, is not agrarian reform. Agrarian reform, as enshrined in the Constitution, with payment in advance and in cash, is an agrarian deal that only interests landowners, radically opposed to the interests of the Brazilian people. That's why the Supra decree is not agrarian reform (STÉDILE, 2005, p. 103-104).

As a result, João Goulart was deposed by the military coup of 1 April 1964 for advocating changes to Brazil's land ownership structure and for having a policy linked to left-wing ideals which, in the context of the Cold War, were becoming forces of opposition to the existing situation in the country.

We therefore stress that the dictatorial coup of 1964 would not have been possible without the intervention of the large landowning class, which is confirmed by the "March of the Family with God for Freedom" organised by the Brazilian Rural Society of São Paulo a few days before Goulart was deposed.

> The 1964 coup, organised by the military and big businessmen, had, among other aims, to prevent the growth of social struggles in the countryside and the political strengthening of rural workers who, for the first time in their history, were entering the political arena en masse (MARTINS, 1984, p. 21).

Thus, according to Sorj (1986, p.23):

> The mobilisations in the period of the João Goulart government acquired characteristics of growing confrontation and polarisation, leading to the unification of a large part of the bourgeoisie around the coup d'état that opposed the reformist movement, thus cutting off the prospects of transforming the land structure through a process of popular mobilisation.

These are, therefore, the struggles that preceded the establishment of the military regime in the country, which acted with particular authoritarianism on the Brazilian agrarian question.

1.2 AGRARIAN REFORM PUBLIC POLICIES

The possibility of a peasant revolution, due to the intense struggles that had been taking place in different regions of the country since the late 1950s, worried the Brazilian government and elite. So the military government of 1964 enacted Law No. 4,504: the Land Statute. It laid down very precise criteria for expropriation; rural properties were thus classified into four categories: latifundia by exploitation, latifundia by extension, minifundia and the rural company, the latter considered to be the ideal category, either because of its size or its form of exploitation, establishing fundamentally capitalist business property. It also states that rural smallholdings and large estates would not be able to fulfil the requirements of the social function of land.

According to Martins (1984, p. 22):

> The Statute opened up access to land when looking at it from the point of view of landowners, but closed off access to land when looking at it from

the point of view of the great mass of landless labourers: agrarian reform would preferably benefit farmers with an entrepreneurial vocation. At the same time, expropriations only took place in the event of conflicts or serious social tension.

Thus, the entrepreneurial vocation expected of rural workers was laid out in the Statute through the creation of the category of rural enterprise. In this respect, Martins (1999, p. 79) states:

> The flexible category of rural enterprise was favoured by the state and escaped the possibility of being included in expropriations. This indicates, in principle, an agrarian reform aimed at economic modernisation and the acceleration of capitalist development in agriculture.

This has led to a concentration of attention on the expropriation of latifundia, which would promote a supposed agrarian reform, but in essence, the imposition of ideal standards and categories seeks to dismantle the struggle of rural workers, acting as a selective relief of tensions in the countryside and maintaining the dominance of the landowners.

The Statute thus reveals its true function: it is an instrument for controlling the social tensions and conflicts generated by this process of expropriation and concentration of property and capital. It is an instrument for encircling and deactivating conflicts, in order to guarantee economic development based on incentives for the progressive and widespread penetration of big business into agriculture and cattle raising. It is an escape valve that operates when social tensions reach the point where they can turn into political tensions. The Statute is at the centre of the government's strategy for the countryside and combines it with other measures to encircle and deactivate conflicts, demands and social struggles (MARTINS, 1985, p. 35):

> The Statute transformed agrarian reform into a topical reform that would serve to alleviate social tensions, a reform subordinated to the process of expanded reproduction of big business capital and also subordinated to a policy essentially of expropriation and concentration of capital. Thus, the government took away the opportunity for regeneration and expanded reproduction of small family farming. Therefore, it can be said that at no point was there any intention, through this law, to carry out a broad and massive reform, compatible with the level of demands of landless or land-poor workers, thus leaving the Brazilian land structure untouched.

Since the military government's intention was not to carry out agrarian reform when the Land Statute was signed:

> [...] some of its implementation was delayed. For example, the GERA (Interministerial Working Group on Agrarian Reform) was only created in

1969, when a mission from the FAO - Food Agricultural Organisation, a UN body, visited Brazil. From this meeting came the suggestion to merge IBRA [Brazilian Institute for Agrarian Reform] and INDA [National Institute for Agrarian Development] into a single organisation to better implement agrarian reform (OLIVEIRA, 2007, p. 121).

The only organisation to which the author refers is INCRA (Instituto Nacional de Colonização e Reforma Agrária (National Colonisation and Agrarian Reform), created by Decree-Law No. 1,110 of 9 July 1970.

In this period, we must emphasise the creation of the SUDAM (Superintendence for the Development of Amazonia), in 1966, and the Banco da Amazônia. SUDAM is the best expression of the transfer of public funds to large agricultural companies, which were establishing themselves in the new agricultural frontiers, such as the Centre-West and Amazonia.

> The creation of the Banco da Amazônia and the Superintendence for the Development of the Amazon (SUDAM) advocated a policy of granting tax incentives to businesspeople, especially from wealthier regions, so that they would no longer pay 50% of their income tax, provided that the money was deposited in that bank to finance development projects in the Amazon, up to 75% of whose capital they held. The investments were orientated towards agriculture, so that a large number of businessmen and companies, especially from the south-east, with no tradition in the sector, became landowners and rural entrepreneurs (MARTINS, 1999, p. 79).

Martins (1984, p.50) reaffirms this:

> Fighting the oligarchy meant expropriating its main power, which was land. The federalisation and militarisation of land in the Amazon became a condition for regional development to leave the hands of the oligarchy, the merchants and traditional landowners, and open up space for big business, giving way to the accumulation of large economic groups, whose scale of operation and interest makes them precisely the effective economic agents of the centralisation of power is the action that gives national scope to the agricultural and industrial products market, the capital market and, what is particularly important in this case, the land market.

Due to the intense peasant mobilisations, the government's understanding at the time was that agrarian reform would be a military issue rather than a social one. The government's offensive on Amazonian lands was great during the dictatorship, however:

> [...] workers had to be brought in so that the plans for "Operation Amazonia" could be implemented, because there was no point in large-scale mining and farming projects in a region where there was a shortage of labour. The alternative was the same one that had long been used in Brazil to make up for the lack of workers: colonisation programmes. [...] The National Integration Programme - PIN - created a highway that started in the Northeast and cut through the Amazon - a route was defined for migration, which was already taking place to Maranhão, Goiás, Pará and

Another strategy of the military government was the promulgation of Decree-Law No. 1.179, of 6 July 1971, which instituted the Programme for the Redistribution of Land and Stimulation of Agroindustry in the North and Northeast (PROTERRA), the aim of which was to "[...] promote easier access to land for men, create better conditions for employment and labour, and foster agroindustry in the areas where SUDAM and SUDENE operate" (OLIVEIRA, 2007, p. 123).

In the 1970s, under Geisel's military government, the entire land policy was linked to the interests of economic policy and the establishment of large farms in pioneer areas, the interests of large economic groups and no longer those of traditional farmers. The few progressive elements in the Land Statute that could have benefited small farmers without land and money were abandoned.

As a result, land policy was redefined in favour of large capitalist companies, while the rural workers' struggle for land grew.

> This period of redefinition of land policy in favour of large capitalist companies, generally industrial, commercial and banking companies, corresponded to a great growth in the workers' struggle for land. Conflicts, despite repression and censorship, multiplied rapidly in all regions of the country, involving not only rural workers, but also indigenous peoples, whose lands were being invaded, with official connivance, on an unprecedented scale. In this context of a growing number of conflicts, the Church [...] was deeply involved in the defence of rural workers [...] (MARTINS, 1984, p. 24).

The persecution of rural workers alienated from the land was great during the years of the dictatorship, but it was during this same period that the first land occupations began to be organised under the influence, above all, of the progressive wing of the Catholic Church, which was resisting the dictatorship. This was the context that led to the emergence of the Indigenous Missionary Council (CIMI) and the Pastoral Land Commission (CPT) in 1975.

To dismantle the Church's action and dissociate it from the struggle for land:

> [...] was direct military interference in conflict situations, through a policy of encirclement and demobilisation of groups of workers involved in the struggle for land, through expropriations and, above all, agreements (MARTINS, 1984, p. 24).

Still aiming to dismantle the most significant land struggle force in the country, the government created the Araguaia-Tocantins Land Executive Group (GETAT) to intervene in the most conflictive area in the country.

GETAT's actions were aimed at reaching agreements, whereby the workers often accepted a smaller piece of land than they were legally entitled to. It was a way of protecting the interests of large landowners and companies, so as to prevent them from losing even more land to the workers, or even the whole of their farms (MARTINS, 1984, p. 24).

We would emphasise that, in this context, the creation of GETAT and the Extraordinary Ministry for Land Affairs represented military intervention in INCRA. The creation of this Ministry practically federalises the land issue (when it institutes federal coordination of state land policies) and places it completely under military control. Martins (1984, p. 25) states that:

> [...] the federalisation of the land question, and its militarisation, makes it easier to neutralise a focus of social and political tensions that was uncomfortable for the military regime and its economic policy. Fundamentally, it centralises decisions on the land problem in the hands of a new minister, eliminating a variety of social groups with common interests [...] in order to implement the agricultural policy of the multinationals and financiers on whose support ambitious government projects depend [...]. The federalisation and militarisation of the land problem will aim to establish, at the very least, an intervention in state land policies, since [...] since the first Republican Constitution, state governments have been responsible for land policy in their respective territories.

On the creation of the Extraordinary Ministry for Land Affairs, Martins (1984, p. 26) states that:

> [...] it turns the problem of land into a barracks problem. It empties the trade union as an instrument for demands and negotiations, and land as the subject of trade union demands. At the same time, it removes the uncomfortable issue of land ownership from the political parties, an issue that none of the parties has so far been able to formulate correctly, divided as they are by the electoral issue and confused formulations about the political project of rural workers, particularly small farmers. The Ministry will try [...] to politically sterilise the debate on land ownership and the struggle for land.

From 1965 to 1981, approximately eight expropriations were carried out per year, despite the fact that there were at least seventy land conflicts in each of these years (OLIVEIRA, 1991).

In 1984, supported by the Pastoral Land Commission, representatives of social movements, rural workers' unions and other organisations met in Cascavel,

Paraná, at the 1st National Meeting of Landless Rural Workers, to found the Landless Rural Workers' Movement.

It should be remembered that this was the period when the country was mobilising for political openness and an end to the military dictatorship and workers' strikes in the cities. The political opening promoted the reorientation of the agrarian question towards the party sphere, thus removing the military character it had had until then.

The new democratic regime had José Sarney at the head of government, a typical representative of Maranhão's clientelism and a large landowner with debatable title deeds, but he had a liberal discourse in favour of reform and social justice.In 1985, the first National Agrarian Reform Plan (I PNRA) was approved and Oliveira (2007, p. 125) says of it:

> The I PNRA already brought setbacks in relation to the Land Statute, such as the article (Article 2, § 2, of Decree No.9 91.766) which states that the expropriation of large estates will be avoided whenever possible. Another point was properties with a large number of tenants and/or partners, where the legal provisions were respected. In this way, the I PNRA already appeared with distortions in relation to the Land Statute. [...] In 1985, with the implementation of the plan, there was a strong struggle between the UDR (Rural Democratic Union), the Sarney government and the landless peasants, squatters, etc. The UDR's aim was to make the implementation of the 1st PNRA unfeasible.

The lack of political will and the prevalence of the interests of landowners, organised in the Rural Democratic Union (UDR), was the reason why, even after two years of implementation, less than ten percent of the goals of the 1st PNRA had been implemented (OLIVEIRA, 2007).

During this period, the Ministry of Agrarian Reform and Agrarian Development (MIRAD), created by the military government, began to organise itself to carry out the necessary expropriations based on the Land Statute; however, before they were sent for presidential signature, the projects to expropriate unproductive farms were rigorously selected, so that many properties were not expropriated.

The 1988 Constitution cancelled out practically all the precarious advances made by the military dictatorship's land legislation. According to Martins (1999, p. 90),

The use of the concepts of "productive property" and "unproductive property" introduced a wide ambiguity in the definition of properties subject to expropriation for land reform, practically cancelling out the relatively more advanced conceptions of the Land Statute.

This change reflected not only the reaction of the large landowners who had organised the Rural Democratic Union [...]. It reflected a fundamental aspect of class alliances in Brazilian history.

The already rhetorical anachronism in the land structure, brought back by the Constitutional Charter, would not have been possible without the action of the UDR, which managed to block the proposal for a broad, general and unrestricted Agrarian Reform in the plenary of the National Congress.

In this way, the ruralists managed to include in the Constitution that productive property could not be expropriated and transferred the setting of rules for the fulfilment of requirements relating to the social function of land to complementary legislation. With the victory of the landowners' land policy, the Sarney government "buried" the 1st PNRA. Firstly, through Provisional Measure 29 of 15/01/1989, it abolished the post of Minister of State for Agrarian Reform and Agrarian Development and transferred MIRAD's responsibilities to the Ministry of Agriculture. Secondly, two months later, Law 7.739 of 20/03/1989 also abolished MIRAD and recreated INCRA through Decree 97.886 of 26/06/1989, linked to the Ministry of Agriculture. The agrarian reform of the "New Republic" ended institutionally in the same way that the military governments had dealt with it, within the Ministry of Agriculture (OLIVEIRA, 2007, p. 128).

As a result, the struggles for access to and permanence on the land came up against, and still come up against, the land-capital alliance, which was and still is against any possibility of substantive changes to the political and social order.

From the 1988 Constitution onwards, policy would be governed not only by the new constitutional order, but also significantly by the adjustment to the globalised economic order, neoliberalism, to which the country has been subject since the 1990s until today. This culminated in offensive actions on the part of the government, both in terms of land regulation and the reaction to the interventions of social movements and actors in the struggle for land.

As far as the struggle for land in the study area is concerned, despite its proximity to Londrina and being a district of that municipality until 1995, Tamarana has a high concentration of agrarian reform settlements negotiated by INCRA, social movement encampments, mainly the MST, Rural Villages and Land Bank Groups, something that doesn't happen very often in the municipality of Londrina. In Tamarana, according to data from the 2010 Demographic Census, the urban resident population is 5,858 and the rural population is 6,404.

As already mentioned, the 1st PNRA came into force during José

Sarney's government, with the aim of massively implementing settlements.

According to Decree No. 91.766 of 10 October 1985, the First National Agrarian Reform Plan (I PNRA), in Article One,

> The National Agrarian Reform Plan - PNRA, presented by the Ministry of Agrarian Reform and Development - MIRAD, is hereby approved for the period 1985/1989, covering 1 (one) million and 400,000 (four hundred thousand) beneficiary families, under the terms of the annex that forms an integral part of this Decree.

It's worth noting that the north of Paraná was the location of the first rural settlement under this plan. This was the Água da Prata settlement, made up of 93 families of squatters who had previously occupied an indigenous area in the municipality of São Jerônimo da Serra, 152 kilometres from Tamarana. However, this was not a recurring occurrence, so the goal of benefiting one million four hundred thousand families was far from being achieved. According to the following graph (Graph 1), 125,376 families were settled between 1985 and 1989, or 874,624 families less than stated in the first article of the I PNRA.

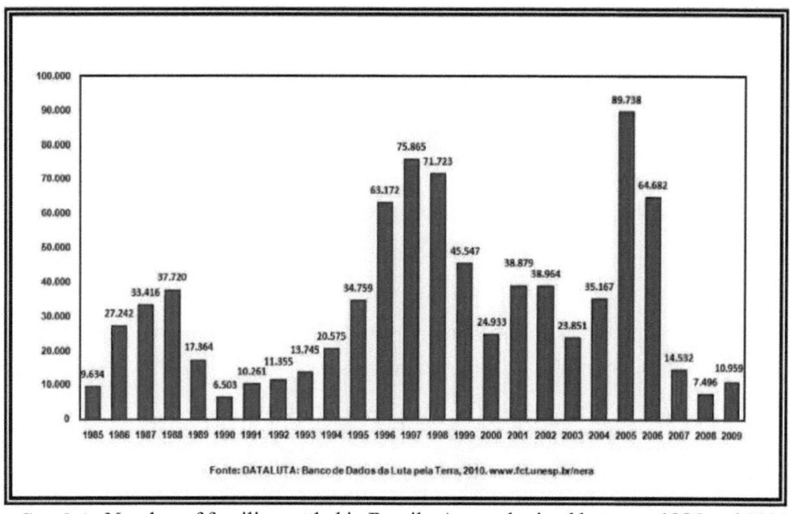

Graph 1 - Number of families settled in Brazil - Areas obtained between 1985 and 2009

As you can see, the years 1995 to 1997 were the most significant in terms of settlements, second only to 2005 and 2006. However, in the last period the methodology for defining settlements was modified, including land regularisation projects[6] , which does not mean opening up access to new families, as expressed in the previous

---

[6] Oliveira (2007) goes into more detail about how Luís Inácio da Silva's government used nomenclatures to inflate land reform figures.

methodology. As such, the first period is a reference in terms of the density of this policy, which coincides with the implementation of eight settlements negotiated by INCRA in the Tamarana region. Table 1 below lists these settlements, as well as the Rural Villages and Land Bank Groups that were subsequently set up in the region in question.

Table 1 - Settlements, Rural Villages and Land Bank Groups in Tamarana - PR

State	Municipality	Location	Type	No. of families	Area (ha)
PR	Tamarana	Cacique	INCRA	11	162
PR	Tamarana	Hope	Earth Bank	8	99,85
PR	Tamarana	Brazil Group	Earth Bank	50	681,78
PR	Tamarana	White Water	Land bank	22	209
PR	Tamarana	Silver Water	INCRA	23	1.651
PR	Tamarana	Mandassaia	INCRA	29	485
PR	Tamarana	New World	INCRA	27	810
PR	Tamarana	Pari-Paró	INCRA	26	-
PR	Tamarana	Reborn 1	Earth Bank	22	100,9
PR	Tamarana	Rebirth II	Earth Bank	18	101,9
PR	Tamarana	Rebirth III	Earth Bank	35	96,8
PR	Tamarana	Sawmill	INCRA	18	381
PR	Tamarana	Treasury	INCRA	24	581
PR	Tamarana	Peasant Union	INCRA	26	547
PR	Tamarana	Rural Village 1	Settlement State	38	-
PR	Tamarana	Vila Rural II - Otávio de Campos Lima	Settlement State	49	-
PR	Tamarana	Penal colony	State Area	11	109

Source: MDA, NERA, organised by the author

Table 1 gives us an idea of the amount of land and the territorial changes they have made to the landscape of the municipality of Tamarana, where 397 plots were distributed, totalling more than 5000 hectares. This place, which originally had a land structure geared towards large estates, began to take on a different shape from that moment on: in the midst of the large extensive livestock properties and some soya, maize and wheat producers, small plots of diversified production and family labour emerged.

This is the context of the implementation of land credit policies, to be discussed in the next chapter.

CHAPTER 2

THE IMPLEMENTATION AND OPERATION OF LAND CREDIT PROGRAMMES IN BRAZIL

The World Bank's land credit policy, known as Market Agrarian Reform, began in Brazil with the "Solidarity Agrarian Reform" programme in Ceará between 1996 and 1997, which was later extended to other states in the Northeast (Maranhão, Pernambuco and Bahia) and the north of Minas Gerais with the "Cédula da Terra" pilot project (19972000). This policy was then expanded to the other states of the federation through the Banco da Terra (1998-2003) and Crédito Fundiário de Combate à Pobreza Rural (2002-2003) programmes. This policy, originally conceived and implemented by the neoliberal government of Fernando Henrique Cardoso (FHC), was based on the creation of a Land Fund to finance the purchase of rural properties whose owners were willing to sell them and landless peasants (or those with little land) interested in acquiring them.

The term market-based agrarian reform arose in the context of criticism levelled by social movements at the Cédula da Terra programme, as a questioning of the World Bank's agrarian policies applied in developing countries. Thus, according to OLIVEIRA (2006, p. 152),

> This expression came to be used in the speeches of intellectuals involved in the peasant struggle for agrarian reform. The expression was then taken up by the World Bank, which placed it at the centre of its political partnership with the Brazilian government, thus seeking to disqualify the critical content of the expression, which was present in its origin.

Despite harsh criticism from social movements and peasant organisations (including the Pastoral Land Commission (Comissão Pastoral da Terra - CPT) and the Landless Rural Workers' Movement (Movimento dos Trabalhadores Rurais Sem-Terra - MST)), these policies continued under Luiz Inácio Lula da Silva, who was the great hope for social movements and civil society interested in reversing the previous government's land reform policy.

With the introduction of the National Land Credit Plan (PNCF) in 2003, expectations of an effective transformation in the country's land reform policy were dashed. This was because the programme maintained the incentive to acquire land through the process of buying and selling on the market, leaving the legal instrument of land expropriation in the background.

The circumstances that led to the proposal of such a policy, coordinated by the World Bank, which was not limited to Brazil, but was directed at countless peripheral countries with pronounced agrarian problems.

The World Bank's (WB) concern with the economic development of peripheral countries such as Brazil, Mexico and South Africa originated in the very creation of the bank in the mid-20th century. The advance of poverty in the world and the financial indebtedness of these countries strengthened the multilateral financial institutions' concern for them. Pereira (2004, p. 13) explains that this concern is due to:

> [...] for two fundamental reasons: on the one hand, the debt crisis was a unique opportunity for the World Bank - a central agent in the elaboration and dissemination of the MRAM[7] - to become, alongside the IMF, the main international financial organisation and thus act as the pivot of neoliberal economic restructuring in Latin America; on the other hand, because the gradual valorisation of market transactions as the main mechanism for land distribution, to the detriment of the expropriationist model, is based on adjustment policies.

The macroeconomic and sectoral adjustment policies advocated by the WB, to which Pereira (2004) refers, are centred on six main axes:

> 1) trade openness, by reducing import tariffs and eliminating non-tariff barriers, with the aim of making markets available, selectively forcing technological modernisation and stimulating exports; 2) deregulation of the domestic market, by reducing state control over prices, incentive mechanisms, exchange and interest rates, etc? with the aim of stimulating competition; 3) financial liberalisation, by reformulating the rules regulating the entry of foreign capital; 4) budgetary and fiscal balance, above all by drastically reducing public spending; 5) deregulation of the private sector; 6) privatisation of industrial companies and the provision of public services (WORLD BANK, 2001a; NAÍM, 1996; SOARES, 1996, apud PEREIRA, 2004, p.13).

The main motivation behind these structural adjustment policies was to ensure that the foreign debt was serviced and to promote the transformation of national economies towards the neoliberal standard that was then gaining strength on the international stage.

In the mid-1990s, the World Bank presented "market-assisted land reform" programmes as an alternative to "rural poverty alleviation" in developing countries. According to the World Bank's documents, these are designed to promote a process in which those interested in buying land can negotiate with those willing to sell, "thus eliminating the usual delays and conflicts and conferring immediate benefits on

---

[7] MRAM - Market Agrarian Reform Model.

small landowners" (BINSWANGER, apud OLIVEIRA, 2006, p. 73).

Furthermore, according to Binswanger (apud OLIVEIRA, 2006), market-assisted land reform could help with the problem of rural poverty in the following way:

> Conventional agrarian reform involves the government expropriating land, with or without compensation, and redistributing it to landless small farmers. This process is often marred by legal disputes and delays that can take decades to resolve. This disproportionately penalises the poor, who spend years before they can occupy the land. Market-assisted land reform aims to facilitate the process in which those interested in buying land can do business with those willing to sell, thus eliminating the usual delays and conferring immediate benefits on small landowners.

The author then points to the justification of slow expropriation processes as a problem that contributes to the permanence of poverty in rural areas. Thus, the WB's "market-assisted land reform" would bring a new solution to old problems, by means of efficient, rapid land reform at low monetary costs.

With this, we can say that this policy denies the instrument legal expropriation, present in the 1964 Land Statute and the 1988 Constitution, so that large estates that do not fulfil their social function, which can be expropriated for the purposes of agrarian reform, are sold, with payment in cash, in currency, at market price. Like this,

> The MRAM is therefore a construct based entirely on criticising and disqualifying another type of land action, considered inevitable and anachronistic in the current phase of capitalism. According to IBRD [International Bank for Reconstruction and Development] theorists, the main difference between the two lies in their nature: while the "traditional" model is seen as "coercive" and "discretionary", since it is based on expropriation, the market model is extolled as "voluntary" and "negotiated" (BURKI; PERRY, 1997, apud PEREIRA, 2006, p. 24).

The "market land reform" model, governed by the World Bank, began in 1994 with experiments in South Africa and Colombia. In Brazil, this model came into operation in 1997, having been implemented under Fernando Henrique Cardoso and followed by the Lula government. This agrarian reform model is part of the WB's offensive to create agrarian policies in line with neoliberal parameters.

The World Bank recognises the need for agrarian reform to deconcentrate land in highly unequal societies, but denies that this needs to happen through expropriation and redistribution, with the state as mediator. For this financial agent, one of the biggest obstacles to economic growth lies in the stance adopted by the state through the use of land as a store of value and legal restrictions on the transfer of ownership and possession rights, thus constituting obstacles to the full development of international policies.

In order to get the land market up and running in the countries and, with it, the "market land reform" model, the WB sells and/or imposes its neoliberal prescription on governments around the world, making successive policies available. According to Rosset (2004, p.17):

> [...] countries start with certain policies, gradually move up the ladder and, in theory, eventually reach other policies. At the moment, we have countries at different stages, and in some the Bank is just beginning to teach them how to take their first steps.

This sequence can be summarised in six policies: land administration, privatisation of public and communal lands, titling through alienable titles, stimulating the land market, land banks - distribution through the market - and, finally, credits for beneficiaries.

Land administration, which the Bank calls the first policy, involves surveying land, cadastre, registration and subsequent demarcation. For the WB, designating this is the beginning of all policies, since its main goal is to create a functioning land market. The Bank's central argument for legitimising the land market is that without the existence of this market, the transfer of land would not take place, so there would be no possibility of poor people acquiring land. This argument can be questioned, because what the WB is really concerned about with regard to the problem of investment in rural areas is:

> Without a market where people can buy and sell land and use it to secure loans or guarantee investors, and people, companies or corporations can obtain title to property rights, according to the Bank, there will be no investment in rural production. Investors demand the security of property rights (ROSSET, 2004, p.19).

With regard to the implementation of the second policy, the WB argues that private land, the latifundia that could be expropriated, would not be enough to meet the demand for land redistribution, so public and communal land should be privatised. This privatisation could be carried out through concessions to companies that agree to invest in rural production or make land available for some kind of agrarian reform compatible with the WB's proposal.

The third policy underpins the other two, and is based on allowing land titles to be alienable, meaning that the land can be sold or used as collateral when applying for credit and therefore lost if the bank loan is not repaid. These titles can also serve as a contribution to a joint venture with a private company, in which the landowner provides his labour and alienable title and the company invests capital. In this case, the risk of expropriation of the peasant is great, because if this business doesn't work, everyone loses, including the peasant, since his land is now alienable.

> Of course, granting a title can, in some cases, fulfil a legitimate and common demand by small farmers to have secure possession of their piece of land. However, a serious problem is that all this is taking place in the context of neoliberal policies - promoted by the World Bank - which undermine the profitability and viability of family farming (ROSSET, 2004, p. 20).

Therefore, this policy of alienable titles, in the neoliberal context, since land policies cannot be evaluated outside of the general economic situation, promotes an increase in land concentration and poverty. This is because the market does not respond to the needs of citizens, but to the interests of capital.

These first three imperatives initiate the functioning of the land market which, combined with the sequencing of land market stimulation, land banks and the provision of credit to beneficiaries, end up constituting the operating framework of the "market land reform" policy.

In line with the interests of the economic elites, the WB argues that land reform, based on the confiscation of property (expropriation), is politically unviable because it generates many conflicts with the rural oligarchy. The Bank also has a guideline prohibiting the purchase of land with its own resources. Based on this assumption, it provides various types of credit funds, created with resources from the country itself or from other donors, so that the states can acquire this land at market prices. There is therefore an orientation for landless rural workers to buy parcels of rural property, or those with little land, generally degraded, unfertile and often overvalued land.

Despite all the capital invested, the problem of land distribution would not be solved due to the corruption that exists within the ruling circles, and even if it didn't exist, the amount of resources needed for its transfer would exceed the capacity of the government and international organisations to carry it out (Rosset, 2004).

We stress that this policy was implemented in Brazil due to internal circumstances which, in the government's view, would be solved by adopting the strategies advocated in the World Bank's "recipe".

The choice of credits for the purchase of land via the market was made in order to reheat the economy, given that investing in land seemed to be the safest option.

It was at this time, more precisely in 1994, that the state, the only possible buyer, opted for such an investment.

This happened because, after price stability was achieved with the implementation of the Real Plan in 1994, the upward trend in land values began to reverse. High interest rates made agricultural costs higher, the overvaluation of the exchange rate reduced the competitiveness of agricultural products abroad, and inflation control eliminated much of the attractiveness of land as a store of value.

In order to fully understand the implementation of "market agrarian reform" projects, it is necessary to refer to the peasant struggles that resurfaced with great force after the last authoritarian regime; as a way of containing this movement, the National Agrarian Reform Plan (PNRA) was launched, which, in its four-year term, fulfilled only 6% of the established target (PAULINO, 2006).

Like this,

> [...] peasants entered the 1990s aware that agrarian reform would not come at the hands of the state. The struggles gained organisation and spread throughout the country, gaining legitimacy and support from broad sectors of society (PAULINO, 2006, p. 304).

It is in this context that we should understand the state's offensive, which in the mid-1990s denied the legal instrument of land expropriation in an attempt to dismantle the struggles for access to and permanence on the land. It was also at this time that the agrarian question ceased to be a matter for the state, as the World Bank acted as the author and majority investor in projects aimed at the reorganisation of Brazilian land.

On the partnership between the Brazilian state and the WB, Paulino (2006, p. 304) explains:

> This partnership is the best indication of the scope and transformative potential that peasant struggles have achieved in this decade, to the point of mobilising an institution that, in terms of principles, is not at all dissimilar to the methods and objectives of the Brazilian government, since it has been a strategic agent for the subjugation of many peoples on the planet to the principles of speculation.

The agrarian question was brought back onto the agenda due to the waves of violence and occupations in the mid-1990s, such as the Corumbiara massacres in August 1995 and the Eldorado de Carajás massacre in April 1996. As a result, Brazil came under international pressure and sought to secure the wishes of the domestic oligarchy through topical measures.

Soon after these events, the Office of the Extraordinary Minister for Land Policy (MEPF) was created in 1996, a body with ministerial powers but without the operational structure of a ministry, and INCRA (the National Institute for Colonisation and Agrarian Reform) was transferred to the control of this new body. A year after its creation, the office was transformed into the Ministry of Agrarian Development (MDA), with a permanent structure.

The Ministry then acted in five directions, as follows: 1) a package of measures - a partial reduction in the final price paid by the state for expropriations, speeding up the time it takes INCRA to take possession of expropriated land, making it more difficult for landowners to evade the expropriation act; 2) using the media to increase the criminalisation of land occupations, prohibiting INCRA from carrying out inspections in occupied areas, making expropriation impossible; 3) the construction of a positive image of the government, through the major media, in relation to land reform, while at the same time creating a negative image of the social movements; 4) the de-federalisation (decentralisation) of land reform policies, transferring to the state level the responsibility for carrying out the entire process of obtaining land and settlements; 5) the implementation of the policy of "market-based land reform" (RAMOS FILHO, 2008).

As a result, Brazil found the ideal situations for the implementation of "market-based land reform", such as the huge demand for land, a tendency for the relative price of rural property to fall, a government strictly aligned with the neoliberal platform and intense social pressure for land; in 1996, the "market-based land reform" policy began in the country.

Under the FHC government, market-based agrarian reform was called the "New Rural World" and centred on three principles that underpin World Bank thinking: the settlement of families as a compensatory social policy; the "stateisation" of settlement action projects, transferring responsibilities inherent to the Union to states and municipalities; and, finally, the replacement of the constitutional instrument of expropriation with propaganda in favour of the "land market".

Through the logic of market-based agrarian reform, the WB promotes the privatisation of land, going against the legal precept that determines expropriation as the main instrument for obtaining land that does not fulfil its social function. This is intended to make peasants integrate into agribusiness, since this is seen as the most efficient production model.

For this integration into agribusiness, the FHC government implemented various projects created by the World Bank, including the Cédula da Terra, initially in the state of Ceará. In 1997, it was extended to Bahia, Minas Gerais, Pernambuco and Maranhão, states chosen for their enormous concentration of poverty, which was to be "alleviated" through market mechanisms.

The Ministry of Agrarian Development justified the implementation of the Cédula da Terra programme as a way to cheapen and speed up access to land and the creation of settlements via the market. One of the aims of the programme was to enable the government to speed up the agrarian reform programme by lowering the cost of land, creating mechanisms that were supposedly more agile and effective than the costly expropriation of land for agrarian reform purposes (SAUER, 2004).

Another project was the Land Bank, approved in 1998; in 2000, the Land Credit to Combat Rural Poverty and, finally, the Settlement Consolidation Programme (which is part of the National Land Credit Programme). As a result, the indebtedness and expropriation of rural labourers was accentuated and land price inflation was reinforced (SAUER, 2004).

The justifications for the implementation of these programmes are based on the assumption that the market and its mechanisms are capable of reducing conflicts and disputes over land, minimising social problems and thus seeking to destructure/de-ideologise movements in the historic struggle for access to land.

The Cardoso government not only encouraged but also invested in market-based land reform, with the main arguments being the slowness of expropriation processes, the overestimation of land values and the high costs of settlements. For the WB, making this model of land reform a success in Brazil was imperative for its dissemination in other countries.

Lula's election as president revived the peasants' hopes. This government was expected not only to reverse the agrarian policy of the land market, but also to make agrarian reform a priority on the national agenda. The political discourse prior to his election placed agrarian reform as an important way of generating jobs, guaranteeing food sovereignty and the basis for a new development model.

However, what we are still seeing today is the continuation of WB policies for rural areas, introduced through the National Agrarian Reform Plan "Peace, Production and Quality of Life in Rural Areas", in which the main goal is the continuation of the Credit Programme to Combat Rural Poverty.

Today, the monetarist view of the agrarian issue, subordinated to the dynamics of the expanded reproduction of agro-industrial capital, reinforces the emptying of the political, social and cultural character given to the issue.

After reflecting on the context of the implementation of land credits, as well as the circumstances that led to their implementation in Brazil, let's turn our attention to how their programmes work.

2.1.1 Cédula da Terra

The Land Credit Programme began to be implemented in Ceará in 1996 through the Solidarity Agrarian Reform Programme - São José Project, run by the Agrarian Development Institute of Ceará (IDACE), in which an association made up of landless rural workers or small landowners had to be formed in order to buy the land.

The purchase of the land was made through a loan between the association and Banco do Nordeste, and the loan had to be paid back after fifteen years, with a four-year grace period, on the outstanding balances, and the Long-Term Interest Rate (TJLP) and the financial agent's rate of return, set at 1% per year on the outstanding balance, would be applied.

In the same year, Law No. 12,614 of 12 August 1996 created the Cédula da Terra, a Pilot Project for Agrarian Reform and Poverty Alleviation, with significant financial support, around 90 million dollars, and intellectual support from the World Bank. This project began to be implemented in 1997 in the Northeast (Ceará, Maranhão, Pernambuco and Bahia) and in the North of Minas Gerais, states chosen for their high concentration of poverty, which was to be "alleviated" through market mechanisms. The target audience would be landless farmers (wage earners, tenants, partners) or those with insufficient land to survive.

With regard to financing the purchase of land, in this programme, in the first phase, the outstanding balance could be amortised over ten years, with a three-year grace period, at interest rates of 4% per year. In the second phase, the interest would be between 2% and 6% per year, with a debt repayment period of 20 years.

According to Oliveira (2006, p.155) the association:

> [...] would have to choose the property to be acquired and discuss the basis of the transaction with the landowner. Next, a proposal for financing the settlement and a request from the owner for a declaration of intention to sell the property would be sent to the state land agency. Once all the documents have been forwarded, the state agency responsible, in this case the land agency, analyses them and then assesses the reasonableness of the

> price of the land on the land market in the area in question. Once this is
> done, a letter of credit is given to the association which, through a state
> financial agent, acquires the property on the market. Contracts are signed
> between the association and the owner of the property, and the property is
> registered in the name of the association and remains its property until the
> debt with the bank is settled. In the event of withdrawal, a replacement can
> be made, as long as the substitute fulfils the eligibility criteria.

Those interested in taking part in the project would have to get together in associations and would be responsible for selecting the area and negotiating the purchase directly with the owners. There would also be a non-repayable investment of R$1,300.00 (one thousand three hundred reais) to get the families settled on the property. However, external influences were decisive in the negotiation process, as research by Sauer (2004) and Oliveira (2006) shows.

The Cédula da Terra programme was conceived and implemented on the basis of market rules, especially with regard to land acquisition. As a rule, the land allocated to the programme is of poor quality and far from the consumer market. The World Bank's argument for this was that the land market in Brazil was incipient and the fact that there were few resources available (the resource limit for each family was US$11,000, including land and infrastructure) would force the acquisition of cheaper areas. However, the fact that the workers participating in this programme were allocated to remote locations, lacking in improvements and infrastructure, with meagre resources to buy the land, equip it with minimal infrastructure and put it to work, only shows the disarticulating nature of this programme.

Limitations such as the implementation of projects in less dynamic regions with less valued, weaker land and serious production restrictions have a direct impact on productive capacity and the conditions for meeting commitments, such as land payments.

The speed with which the Cédula da Terra project was extended to the whole country was related to pressure from social movements to speed up expropriations for land reform and the constant criticism the programme was receiving. According to Medeiros (2003), the creation of the Banco da Terra, a proposal to expand the Cédula da Terra, corresponded to the demand of landowners who had always opposed agrarian reform programmes based on the expropriation of unproductive properties and the payment of agrarian debt bonds.

2.1.2 Earth Bank

Despite harsh criticism from social movements and peasant

organisations (including the Pastoral Land Commission (CPT) and the Landless Rural Workers' Movement (MST)), the Land Bank or Land and Agrarian Reform Fund (Fundo de Terras e da Reforma Agrária) was established in 1998 by Complementary Law 93/1998.

The purpose of these loans was to finance small farmers - organised in associations, cooperatives and condominiums - to buy rural properties and set up basic infrastructure. Instead of promoting the expropriation of unproductive land, it was negotiated by the farmers and acquired at market prices with Banco da Terra funds, and mortgaged until the loan was paid off.

The project has been the target of much criticism and controversy since its creation, the main criticism being linked to the substitutive role given to the Land Bank in relation to redistributive agrarian reform. In this way, the state gave up its role as intervener and left the responsibility for land reform to the market. According to Gérson Teixeira, president of the Brazilian Agrarian Reform Association (ABRA):

> FHC thought that the market would have the power to democratise the structure of land ownership, but he left the power to do so to the latifundia themselves and to the landowners. In this way, the Land Bank cannot function as an agent of agrarian reform (MAIOR; SALVO, 2003, p.1).

The Land Bank Programme is structured as illustrated in Figure 2, which shows the national, state, regional and municipal levels responsible for its implementation.

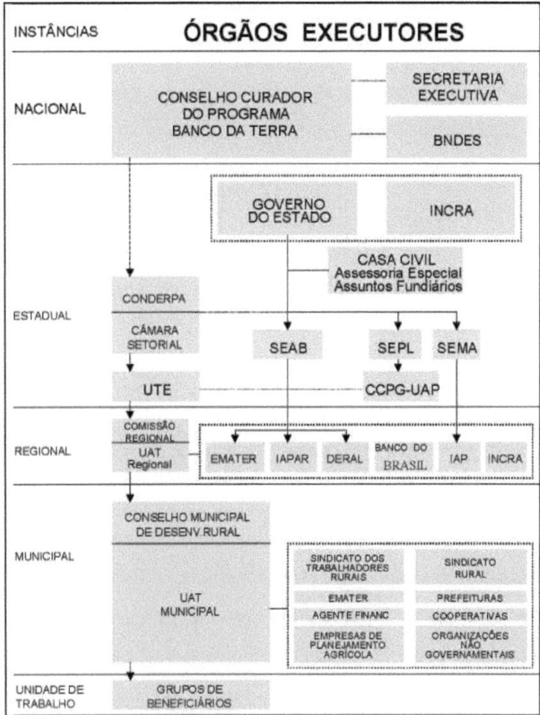

Figure 2 - How the Banco da Terra Programme works

Source: Banco da Terra Operations Manual (2000)

According to the Banco da Terra Operations Manual, the Programme's specific objectives are,

> a) to provide the necessary means for the acquisition of agricultural land and the establishment of infrastructure for rural workers, landless farmers and farmers with insufficient land to maintain their families;
> b) rationalising the efforts and use of financial resources from the federal, state and municipal governments, according to the needs of the beneficiary groups;
> c) provide the areas of the settlements or their influence with the infrastructure and services needed to improve agricultural business and social conditions.
> d) promote integrated action between the groups of beneficiaries and the other mobilising agents involved in the Programme, particularly the municipal government, encouraging local decision-making based on the municipal development plan;
> e) contribute to providing jobs and keeping families in rural areas;
> f) encourage and support the organisation of family farmers to achieve their goals together;

However, in our view, few of these objectives were met, as we will show through our visits to the groups studied.

The Banco da Terra had some differences from the Cédula de Terra programme. Landless rural workers, small farmers and their children could take part in the programme, and land could be acquired through the organisation of associations or individually. The financing could be paid back in up to twenty years with a three-year grace period and subsidised interest rates ranging from 2% for the poorest areas to 6% per year. However, the interest rate could vary according to the value of the individual land financing.

Thus, according to Banco da Terra's Operations Manual, the following effective interest rates apply, depending on the amount of financing per beneficiary:

> a) up to R$ 15,000.00 (fifteen thousand reais): 6% p.a. (six per cent per year).
> b) above R$ 15,000.00 (fifteen thousand reais) and up to R$ 30,000.00 (thirty thousand reais): 8% p.a. (eight per cent per year).
> c) above R$ 30,000.00 (thirty thousand reais) and up to R$ 40,000.00 (forty thousand reais): 10% p.a. (ten per cent per year).

The inadequacy of the model, coupled with the discovery of various irregularities, led to the suspension of the Banco da Terra programme in February 2003.

Banco da Terra ran from 1999 to 2002 (serving 34,478 families) in Brazil, a period of government transition that culminated in the victory of candidate Luís Inácio Lula da Silva, elected by a political coalition led by the PT, with the support of social movements and peasant organisations fighting for land reform. The following table summarises the results of the programme.

Table 1 - Balance of Banco da Terra operations between 1999 and 2003

State	Contracts	Municipalities	Families	Area (ha)	Value	Average value per hectare
Alagoas	24	21	656	10.476	R$ 13.118.448	R$1 .252,23
Holy Spirit	21	16	586	5.760	R$ 12.416.380	R$2 .155,62
Goiás	27	21	2.259	33.744	R$ 45.397.198	R$1 .345,34
Maranhão	1	1	33	827	R$ 120.196,00	R$145 ,46
Minas Gerais	94	76	2.534	167.400	R$54 .385.844	R$324 ,88
Mato Grosso do Sul	17	13	1.212	14.869	R$ 22.847.834	R$1 .536,60
Mato Grosso	41	30	3.214	64.766	R$ 57.115.718	R$88 ,44
Paraíba	68	58	452	25.010	R$ 13.866.413	R$554 ,43
Pernambuco	4	3	121	4.112	R$ 2.196.675	R$534 ,21
Piauí	34	30	1.436	41.458	R$ 16.454.551	R$396 ,89
Paraná	123	88	2.160	24.210	R$ 64.524.797	R$2 .665,21
Rio de Janeiro	11	9	349	4.371	R$ 8.349.162	R$1 .910,12
Rio Grande do Norte	19	16	496	10.224,00	R$ 7.701.461	R$753 ,24
Rio Grande do Sul	1.251	437	10.239	119.301	R$ 212.520.538	R$1 .781,38
Santa Catarina	843	264	4.685	75.426	R$ 140.294.114	R$1 .860,02
Sergipe	24	17	1.024	11.325	R$ 17.151.402	R$1 .514,47
São Paulo	71	58	2.093	14.189	R$ 63.910.812	R$4 .504,25
Tocantins	11	9	382	9.567	R$ 487.658	R$506 ,70
Total	**2.684**	**1.160**	**34.478**	**1.218.035**	**R$ 757.219.302**	**R$1 .323,86**

Source: MDA (2007)

Table 1 shows that the Programme's most intense activity was in the South, which accounted for 70% of the financing contracts. This is due to a number of factors, most notably the saturation of dismemberment due to the historical settlement process, in which small properties were numerically important. Also contributing to this performance was the specific case of the state of Rio Grande do Sul, where individual contracts were a widely used expedient and, finally, the distinctive role of local authorities, trade unions and town halls in making Banco da Terra a reality. Another fact worth highlighting is the price of land, as in the state of São Paulo, which is up to 50 times higher than in Mato Grosso. This is because the price is influenced by the quality of the soil, as well as the presence of infrastructure, which means that the areas of plots in places where land is more expensive are smaller.

However, before the end of President Fernando Henrique Cardoso's administration (1998 - 2002), the Ministry of Agrarian Development, under former minister Raul Jungmann, launched a new market-based land access programme to replace the Banco da Terra, the Crédito Fundiário e Combate à Pobreza Rural (CFCP) project.According to Medeiros (2003), CONTAG's (National Confederation of Agriculture) support for the programme decisively influenced the approval of the loan with the World Bank, which shifted support from the Land Bank Programme to the Land Credit Programme to Combat Rural Poverty. No less important than this influence was the open opposition by social movements and peasant organisations via the National Forum for Agrarian Reform and Justice in the Countryside.

2.1.3 Rural Poverty Credit

The Land Credit and Fight against Rural Poverty Project maintained the same proposal as its predecessor programmes, i.e. to alleviate rural poverty in the Brazilian countryside. Financed by the federal government and the World Bank, it brought as a novelty the participation of CONTAG in its creation and implementation.

The CFCP covered all the states in the Northeast and South regions, as well as Minas Gerais and Espírito Santo[8] . Its proposal allowed for individualised credit (without the right to productive resources) and financing was maintained with a term of up to twenty years, with a three-year grace period, fixed interest of 6% per year, with a 50% rebate, applicable to financial charges if the instalments were paid on time, as well

---

[8] According to the Ministry of Agrarian Development, 77,000 families benefited from the Land Credit Programme to Combat Rural Poverty between 2003 and 2009. Available at:<http://www.mda.gov.br>. Accessed on 16 June 2010.

as exemption from monetary correction.

The Cédula da Terra, Banco da Terra and Crédito Fundiário e Combate à Pobreza Rural projects have in common the logic of valorising the private appropriation of land. Although they look like the same "market agrarian reform" policy, they differ in structure and design. The permanence and changes, the agility in development and the mobility in changing programmes, reveal the struggle for land in Brazilian society, between social movements and the policies of the Fernando Henrique Cardoso government. During Lula's administration, these struggles continued; however, they had little impact, not because the government really bothered to treat land reform as a necessity, but because it implemented land credit programmes wherever there was tension over land, as can be seen in Graph 2.

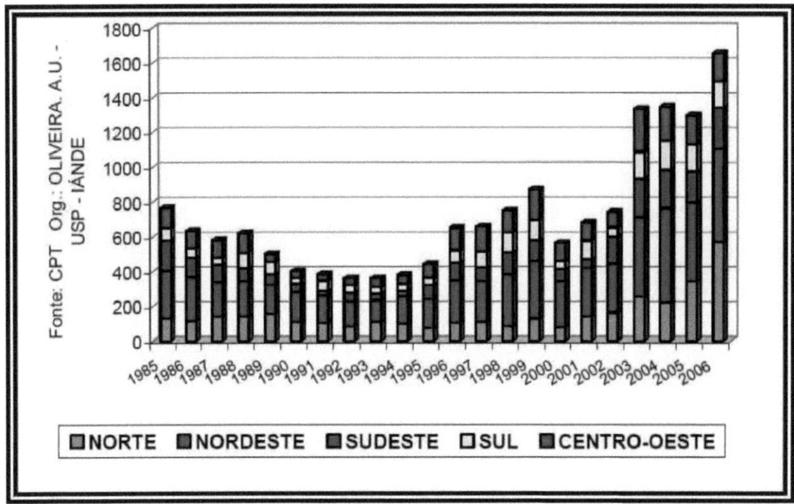

Graph 2 - Conflicts in the countryside in Brazil 1985 to 2006

Source: Oliveira (2007, p. 154)

2.1.4 National Land Credit Plan

The PNCF was presented as a historical demand by trade union organisations such as CONTAG and FETRAF-Sul (Federation of Workers in Family Agriculture in the Southern Region), for a credit programme to complement land expropriation and as an effective mechanism for social participation and control. It defined the non-acquisition of unproductive areas larger than 15 fiscal modules and social participation through rural workers and their community organisations, the

Municipal Council for Sustainable Rural Development (CMDRS), the State Council for Sustainable Rural Development (CEDRS), the State Technical Unit (UTE), the National Council for Sustainable Rural Development (CONDRAF), the federations of workers in agriculture and family farming and their unions.

In this way, the programme would not repeat the problems that occurred in previous experiences, nor would it be part of any "market land reform", fundamentally because it would not buy areas that could be expropriated and because it would have a participatory structure, from preparation and management to execution.

The PNCF is subdivided into three lines of financing: Combating Rural Poverty, Our First Land and Consolidating Family Farming, targeting landless rural workers, small rural landowners with precarious access to land and small landowners, including young people and the elderly. It also had as a principle the "autonomy" of the communities, i.e. the association should self-select its participants, choose the property to be bought, as well as negotiate the price of the land with the owner, and also how to spend the resources intended for community investments, in addition to managing its own forms of organisation and production, among other activities.

Financing for the purchase of the land would come from repayable resources from the Land Fund and financing for community investments from non-repayable resources, mainly from the contract with the World Bank. These investments are divided into three types, namely: a) productive investments (subsistence and production gardens, pasture formation, facilities, soil conservation, costing of first crops, purchase of animals); b) basic infrastructure investments (housing, water supply, electrification, roads within the property, etc.) and c) investment for the formation of savings by families or communities (bank savings or investment funds, formation of working capital, etc.).

The farmers taking part in this programme also have non-refundable initial support for installation of up to R$2,400.00 (two thousand four hundred reais) and, in the case of properties located in the semi-arid region, an additional R$2,000.00 (two thousand reais) per family to cover water security costs. The farmers' contribution would be at least 10% of the total investment, in the form of labour, materials or money, but it was basically in the form of labour. The source of funds for the programme was the Land Fund (formerly Banco da Terra), created by Complementary Law 93/98 and regulated by Decree 4.892 of 25 November 2003, and the World Bank (OLIVEIRA, 2006).

The PNCF has fuelled the discontent of social movements and peasant representatives in the struggle for agrarian reform with the agrarian policy of the Lula da Silva government. This is because the programme maintained the incentive to acquire land through the process of buying and selling on the market, leaving aside the constitutional instrument of expropriation, and introduced the strategy of managing the Land Fund on a long-term basis (thirty years), signalling a long time for payment. As a result, the newly settled peasants will have to spend more on paying back the loans.

When we analyse the implementation of "market-based agrarian reform" in Brazil, as well as the way its programmes have worked, we can say that public policies for specific segments tend to privilege and strengthen thc latifundia; and agrarian reform thus emerges for the government as a response to pressure from social movements fighting for land.

In this way, the World Bank's agrarian reform policy has neglected the peasant struggle for agrarian reform and maintained an agrarian policy that values the market, rather than the state, as the legal instrument for access to land.

We therefore believe that the market is not a mechanism capable of democratising Brazil's land concentration or promoting any kind of socially just agrarian reform.

So let's move on to the evidence of this contradictory process of peasant re-creation, in which land credit projects are a good example because, despite favouring the land market, which makes the peasant condition so vulnerable, they open gaps for its re-creation, obviously in a percentage that is negligible compared to the total number of beneficiaries.

CHAPTER 3

CHARACTERISATION OF THE STUDY AREA

As seen above, Tamarana is one of the municipalities in the north of Paraná where land distribution policy was an important expedient, especially at the end of the 1990s, which must be attributed to the ability of organised peasant movements to put the demand for land on the agenda.

The geography of the municipality also played a part in this, especially the existence of a stock of titled land with vices of origin, which obviously encouraged farmers to have it expropriated by INCRA. Add to this the topography and soil conditions that make some of the land have a low market price, making it attractive for public acquisition under the terms already presented in this work.

These are the reasons why there are also a significant number of land credit projects there, which led us to the proposed cut-off, also adding the Banco da Terra projects in Londrina and Tamarana, as shown in Figure 3, whose geographical characteristics (type of soil, climate, vegetation, among others) are similar, thus providing some parameters for comparison.The Banco da Terra groups analysed in this research are shown in Table 2 and Table 3, along with their location and the number of families living on the properties.

Chart 2 - Land Bank groups in Londrina

Location	Group Name	Number of families	Year of Incorporation
Londrina (Lerroville)	Alto Alegre Group	47	2008
Londrina (São Luiz)	Akolá Group	42	2001
Londrina (Warta)	King of Lettuce Group	06	2001

Source: EMATER - Londrina. Organised by the author.

The districts in which the Bank Groups are located are

The Land Groups of Londrina are shown in Figure 4.

da Terra in the municipality of Tamarana, Table 3 gives details of its constitution.

Chart 3 - Land Bank groups in Tamarana

Location	Group Name	Number of Families	Year of Incorporation
Tamarana	Brazil Group	50	2001
Tamarana	Hope Group	08	2001
Tamarana	Renaissance Group 1	22	2003
Tamarana	Grupo Renascer II	18	2003
Tamarana	Grupo Renascer III	17	2003

Source: EMATER - Tamarana. Organised by the author.

Figure 3 below shows that the Banco da Terra groups are often located on low-fertility soils, often shallow, with rocky outcrops, such as the eutrophic litholic A soils found in the Renascer I, II and III Groups. In the Brazil, Esperança and Alto Alegre Groups, the predominant soil is the dystrophic purple latosol A, an acidic soil with a high concentration of clay and low water storage capacity. Due to the steep slopes, especially in the Brazil Group, it is only suitable for growing vegetables, some cereals such as maize, soya and wheat, or livestock.

Fazenda Akolá and Rei do Alface are located in a predominantly eutrophic A structured purple soil, which is a highly fertile, deep, alkaline soil with little or no rocky outcrops.

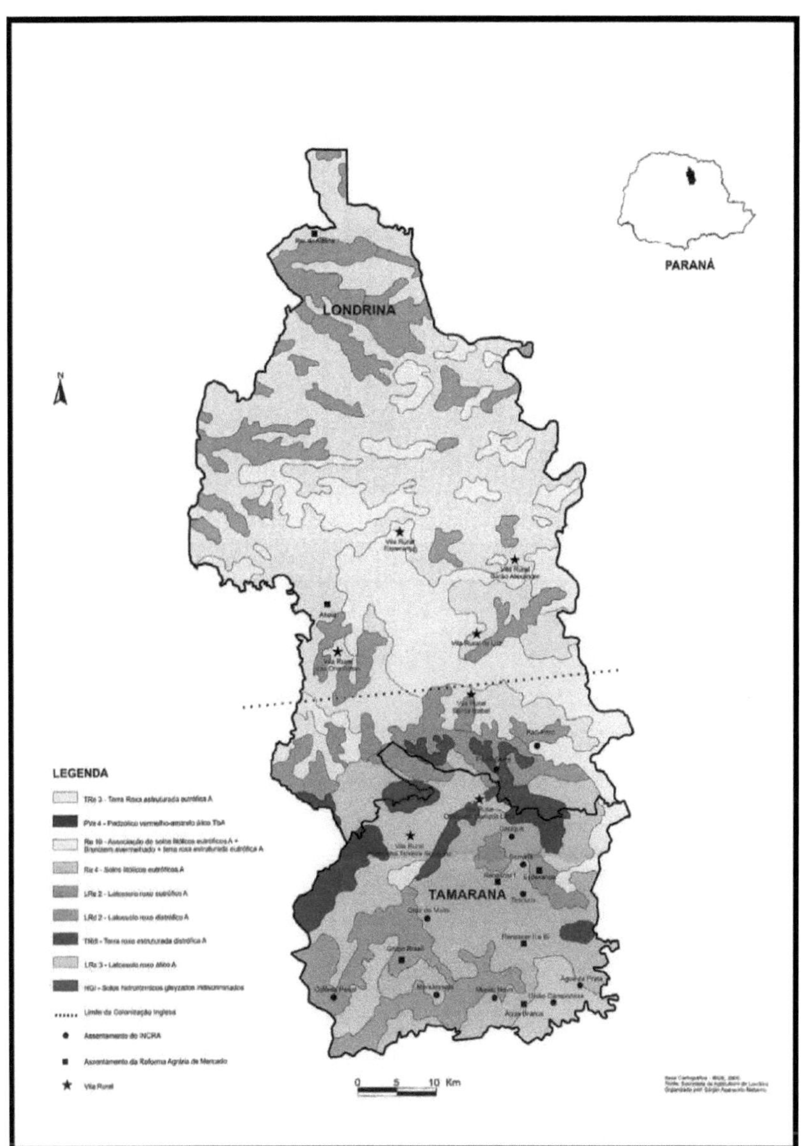

Figure 3 - Soil map and location of Settlements, Rural Villages and
Land Bank Groups
in the municipalities of Londrina and Tamarana
Source: Nabarro (2010, p. 114)

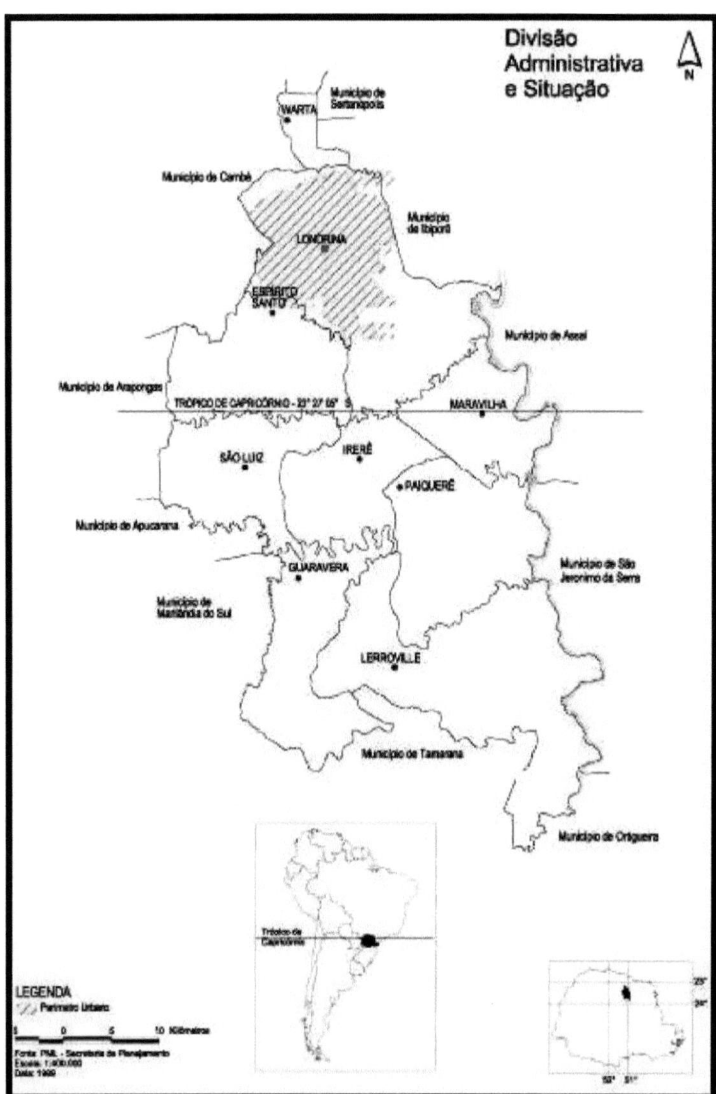

Figure 4 - Londrina - Administrative Division
Source: Londrina Planning Department.

3.1.1 Akolá Farm

Fazenda Akolá is located in the district of São Luiz, in the municipality of Londrina/PR, with a total area of 485.6 hectares (ha), 152 hectares of which are earmarked for the (collective) legal reserve. The remaining 333.6 hectares are home to 42 families (181 people), each with a 7.94 hectare plot.

The farm was acquired through financing from the Federal Land Credit

Programme - Land and Agrarian Reform Fund (Banco da Terra) in 2000. The purchase price of the property was R$1,465,213.40 and the financing price was R$1,680,000.00, with the total amount of the financing for each "beneficiary" being R$40,000.00. The programme offers a grace period of three years for payments to begin, and the loan lasts 17 years.

The purchase price of the property (R$ 1,465,213.40) is made up of the sum of the values comprising the price of the bare land R$ 1,384,743.80, improvements/equipment[9] R$ 20,000.00; notary costs R$ 47,945.20 and surveying and area measurement costs R$ 12,524.40.

In order to analyse what the cost of the land represented in the debt taken on by the beneficiaries, we first looked at the base price per hectare provided by the Paraná Department of Agriculture and Supply (SEAB) and the Department of Rural Economy (DERAL) in Londrina for mechanisable and non-mechanisable land, to see whether or not the property had been overvalued. The base price for mechanisable and non-mechanisable land in 2000 (the year the property was bought to set up Banco da Terra - Fazenda Akolá) was R$2,893.00[10] p/ha and R$1,859.00[11] p/ha respectively. However, the price paid per hectare, according to the financing proposal made by EMATER, was R$2,851.61.

Having clarified these figures, let's do a simple exercise: if we know that, on Fazenda Akolá, the area of non-mechanisable land is comprised only of the legal reserve (152 hectares), this leaves 333.6 hectares of non-mechanisable land.
mechanisable land. If we multiply the mechanisable area by the base price, we get a value of R$965,104.80; if we repeat the same reasoning for the non-mechanisable land, we get a value of R$221,280.00. If we add the two values together, we arrive at the real value of the property of R$1,176,784.80, which represents a difference of R$208,359.00 on the price paid by the Banco da Terra Programme to the owner of the area.

Assuming an unrealistic situation in which only the total area of the farm is considered to be mechanisable and multiplied by the base price, only then would there be a higher price than the one paid to the owner, of R$1,404,840.80, i.e. a difference of R$20,097.00.

However, the questions don't just rest on the price paid for the land, as

---

[9] The existing improvements on the property were two masonry houses, one with an area of 60m² (R$8,000.00) and the other with 48m² (R$3,500.00) in good condition, according to the financing proposal made by EMATER technicians, a well with a water reservoir (R$5,000.00) and electricity (R$4,500.00).
[10] The price refers to a hectare of mechanisable purple land in Londrina in 2000.

it is well known that the properties included in the Banco da Terra programme would be sold at market price, not at a pre-established price. The question also arises as to how the total value of the financing (R$1,680,000.00) was arrived at: the price of the property (R$1,465,213.40) and 54% of the costs of the infrastructure project (R$214,786.98) were added together. This project included the construction of forty masonry houses at a total cost of R$200,000.00, rural electrification (R$87,120.00), basic sanitation (R$46,000.00), the construction of access roads (R$13,954.00) and 42 granaries (R$42,000.00).

The largest amount allocated to the "beneficiaries" for the completion of the infrastructure project was R$160,000.00 for the construction of the houses, about R$4,000.00 for each house and the rest (R$40,000.00) for the group. Thus, each house was to be built with R$5,000.00, based on a borderline situation in which the person who activated the financing via Banco da Terra would only have this money to carry out the infrastructure of their plot.

With these figures, we want to show the unacceptable situation in which a programme like Banco da Terra and other land credit programmes are aimed solely at financing the purchase of land, in a country that has 216,328,597[11] [12] hectares of unproductive land, declared by its own holders to INCRA, and 496,901,265[13] hectares of vacant land, in other words, an area of 713,229,862[14] hectares that could be expropriated in the social interest for the purposes of land reform. According to Article 184 of the 1988 Constitution:

> Art. 184: The Federal Government is responsible for expropriating rural property that is not fulfilling its social function for the purposes of agrarian reform, by means of prior and fair compensation in agrarian debt securities, with a real value preservation clause, redeemable within a period of up to twenty years, from the second year of their issue, and whose use will be defined by law.

In the area studied, the situation is no different: in Paraná, 20% (3,956,951ha[15]) of the total area (19,927,978ha[16]) is wasteland, and in Londrina, 18% (31,047ha).[17]

It is also worth questioning whether it is possible for a family with few

---

[11] The price refers to a hectare of non-mechanisable purple land in Londrina in 2000.
[12] OLIVEIRA (2010 s/p.)
[13] OLIVEIRA (2010 s/p.)
[14] OLIVEIRA (2010 s/p.)
[15] OLIVEIRA (2010 s/p.)
[16] OLIVEIRA (2010 s/p.)
[17] OLIVEIRA (2010 s/p.)

monetary resources, which joins a land credit programme like Banco da Terra, to be able to immediately produce in order to provide for their livelihood and social reproduction. How would it be possible for a programme that only aims to finance a rural property, since none of Banco da Terra's resources are earmarked for productive projects, to be able to promote the consolidation of peasants and their recreation?

In order for the production project to be put into practice, the producers have to activate another credit programme, PRONAF, for both investment and funding. At Fazenda Akolá, the initial production project cost R$399,000.00, and the crops chosen were densified coffee, maize, cassava, yam and lettuce. With the exception of lettuce, all the crops were funded by PRONAF A and D. According to data from the Paraná Institute for Technical Assistance and Rural Extension (EMATER), the average annual per capita income is R$6,738.00 (R$561.50 per month), 94% of which comes from agricultural production and 6% from other sources of income. The predominant production system on the farm is olericulture, comprising crops with an average area of 5.69 hectares of yams, sweet potatoes, cassava, yams, green corn, beetroot, carrots and aubergines; and perennial crops with an average area of 3.74 hectares of oranges and densified coffee. This data was presented on 3 December 2009 at the "Rural Community Walk" event, as shown in Photo 1, in which we can see the president of the Akolá Farm Producers' Association talking to the event's participants.

Photo 1 - Walk in the Rural Community - Akolá Farm

Source: EMATER - Londrina, Dec. 2009.

According to EMATER extension worker Paulo Mrtvi (2009, p. 1), the main objective of the event was:

> [...] to highlight the importance of the process of organisation, of collective work, especially when it comes to selling production and buying inputs. At the time, producer Agnaldo Bispo illustrated this idea well: through the association he got R$14.00 for a box of potatoes; if he were to sell them to middlemen, he says, he wouldn't get more than R$5.00.

This event was particularly valuable for collecting data on Fazenda Akolá, and was also the source for the tables and graphs presented below.[18]

Table 4 shows land use and agricultural aspects and Graph 2 shows the amount of machinery or implements on Fazenda Akolá.

Table 4 - Land use

Ploughing	Area (ha)	Productivity kg/ha	No. of Producers
Coffee	2,42	1.200	03
Maize	80,0	7.000	30
Cassava	150,0	18.000	42
Cará	170,0	18.500	42
Vegetables	8,0	20.000	03
Virgin Forest	152		42
Pasture	4,33	06 heads	06
Roads and paths	2,01	-	-
Other	12,3		
Sweet potatoes	121,0	20.000	40

Source: EMATER - Londrina. Dec. 2009.

Analysing the table above, we can see that scale crops, such as maize, have little significance in the production of Akolá's farmers. The explanation lies in the fact that the useful area of each plot is limited - only 7.26 hectares. Another factor that makes it unfeasible to grow crops on a large scale is the high investment required in production and the low profitability, since an extensive crop such as grains will only be highly profitable when produced on a large scale.

Table 5 shows maize productivity in Brazil, Paraná, Londrina and Fazenda Akolá. It is clear from the table that, even with a smaller area, the farm's productivity is high compared to national, state and municipal productivity.

Despite this differential, in terms of productivity, this is not enough to guarantee the income needed to pay off the debt and other family recreation needs.

---

[18] The data was kindly provided by the EMATER - Londrina technician responsible for the event.

Nevertheless, this data leads us to question the national productivity index, which should be updated periodically, in accordance with Law 8.629 of 25 February 1993, as a way of ensuring compliance with the social function of property. However, it was only in 2005 that the MDA, together with INCRA, sent the first proposal for an ordinance to update the productivity indexes.

Thus, according to Paulino (2009b):

> Although this happened in April 2005, the concrete result was the articulation of the rural sector within the government itself, preventing the decree from being issued. The Minister of Agriculture, Livestock and Supply of that administration, Roberto Rodrigues, became the spokesperson and defender of the sector, creating a coalition of forces that has so far prevented the indexes from being updated. This is the clearest evidence that, in Brazil, land surrounded by large estates does not fulfil its social function, even if there is a segment of efficient and competitive agribusinesses within it. However, it is certainly not significant among large properties, because if it were, there would be no obstacle to revising the indices.

The indices currently used date back to 1980 and are based on indicators of crop productivity, the technical level of agriculture and livestock per hectare, according to data from the 1975 agricultural census, data that was used to implement the First National Agrarian Reform Plan during José Sarney's government.

The proposal for the new calculation, submitted in 2005 and intensely debated by the rural caucus in the National Congress, defending its own interests and those of those who maintain control of unproductive properties, creates indices that vary from region to region and from crop to crop. On average, the new proposal increases the current productivity index by 100 per cent; however, because the current indices have long since been surpassed and because of the increase in agricultural productivity driven by new technologies and techniques, less than 7 per cent of Brazilian rural properties would be affected, in Paraná around 2 per cent.

In August 2009, President Luiz Inácio Lula da Silva had set a deadline of 15 days for the publication of an inter-ministerial decree revising the indices, but once again pressure from the rural caucus blocked the decree.

According to Sciarra (2010):

> To put pressure on the government to back down, the ruralists passed bills in the House and Senate that would remove the Executive's power to revise the indices. Even so, the government's technical team considered the update to be quite reasonable, because it would be restricted to a few municipalities in the country. In the general average for the main crops, 90 per cent of the municipalities would have new indices lower than or equal

to the historical averages calculated by the IBGE over the last ten years. The proposal from the working group of the Ministries of Agriculture and Agrarian Development would change the minimum rates in only 369 of the 4,842 municipalities where soya is produced in the country. In other words, it would change the requirements in only 7.6 per cent of these municipalities. The proposal would include only 1.2 per cent of the 5,512 corn-producing municipalities (640), 11 per cent of the 4,442 orange-growing localities (488) and 2.7 per cent of sugar cane (146). The filtering carried out by the government took IBGE's average productivity into account.

Table 5 - Productivity - Maize

Location	Year	Area (ha)	Production (tonnes)	Productivity (kg/ha)
Brazil	2007/2008	14.444.582	58.933.347	4.080
Paraná	2007/2008	2.972.248	15.602.601	5.249
London	2007/2008	6.000	47.520	7.920
Akolá Farm		80	-	7.000

Source: SEAB - DERAL/IBGE (2010) organised by the author

Although the maize productivity index at Fazenda Akolá is higher than in Paraná and Brazil, as shown in Table 5, producing maize for sale is unfeasible on small areas, as will be shown below when we look at production and marketing costs.

The estimated cost of corn production in August 2010, according to data from DERAL and SEAB, is around R$20.00 per 60kg bag. However, the price paid to producers for a sack of corn on 27 August 2010, according to the SEAB/DERAL Agricultural Market Information System (SIMA), was R$15.26 for Londrina; therefore, the selling price of the product does not cover the costs of its production.

The high cost of maize production is related to fertiliser expenses, at around R$327.50[19] per hectare, pesticides (R$33.54 p/ha) and seeds (R$154.62 p/ha). This converges with the bet on transgenics. By using genetically modified seeds, especially maize, the aim is to reduce the cost of pesticides, since the transgenic seed varieties produced and commercialised in Brazil are resistant to insects. Therefore, it should not be inferred that productivity in areas with transgenic maize is higher than in areas with conventional plants, but there would supposedly be a reduction in pesticide costs.

Even the high production potential of seed varieties and hybrids requires intensive use of pesticides and fertilisers in the production cycle, which means high investments.

---

[19] Estimated cost of maize production in May 2010, by SEAB-DERAL.

Paulino (2009b) notes that "the more economically vulnerable farmers are, the proportionally lower their crop yields will be, even if they buy the best seeds"

Thus, the use of transgenic seeds as an alternative to the high costs of maize production is not in line with the reality of the producers surveyed, as the average price of transgenic seeds is R$380.00 and conventional seeds around R$70.00 to R$200.00.

We should also point out that Fazenda Akolá has a university research and extension project on the production, conservation and genetic improvement of its own seeds, particularly maize. Set up in 2006, it involves three lecturers from the State University of Londrina, respectively from the Geosciences Department and the General Biology Department, and an extension worker from EMATER.

According to Paulino (2009b):

> The principle that has driven this experiment is the possibility of combining tradition with cutting-edge scientific knowledge, as is the case with the seeds project. As we know, access to seeds is one of the inalienable patrimonies of humanity, which for millennia has selected and conserved them with a view to subsequent harvests. [...] We know that farmers stopped producing their own seeds when they realised that they were no longer competitive with commercial seeds. [...] In short, the project works to train farmers in the production and participatory genetic improvement of their own seeds, making it compatible with their production systems and the resources available on the property, while at the same time trying to recover knowledge, traditions and skills that were partially or totally lost after the emergence of seed companies.

As far as this project is concerned, we would like to point out that only a few farmers have signed up to the proposal, particularly those who need maize for their own consumption, as the area available does not allow for large-scale production of this cereal.

In fact, it is precisely this material situation, i.e. the size of the plots, that contributes to understanding the data presented in Graph 3, where we can see that the number of tractors is relatively low, while animal traction is present on almost all the plots.

Graph 3 below shows that all the producers have general implements, such as a hoe, a manual seedling planter and basic tools to enable the land to produce. We also see that more than 90 per cent of producers have animal traction.

The number of tractors, as already mentioned, is low, and 20% of them were bought through the Paraná state government's programme called Trator Solidário

(Solidary Tractor), which allows farmers, provided they "benefit" from PRONAF C, D and E, to buy 50 HP tractors for R$40,100.00[20] and 75 HP tractors, also for R$47,250.00[21] , as well as new agricultural implements and equipment. Although this programme appears to be easy, it is restricted to some farmers and fails to benefit many others who do not meet the requirements of PRONAF C, D and E, as shown in Table 6 below.

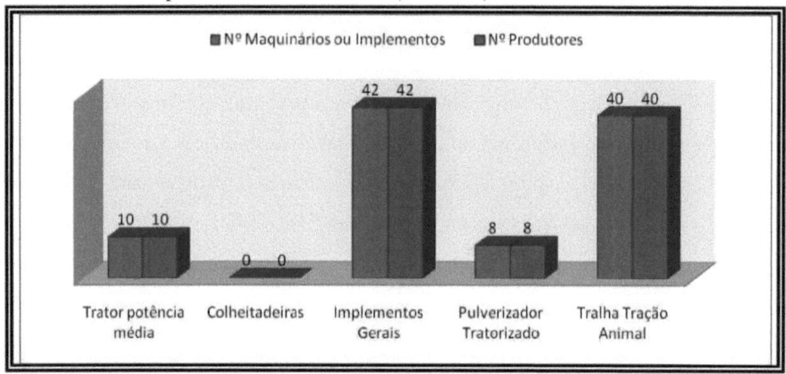

Graph 3 - Akolá Farm - Machinery and Implements
Source: EMATER - Londrina. Dec. (2009).

Chart 6 - PRONAF Framework and Criteria

FRAMEWORK	CRITERIA
GROUP "A"	- Agrarian Reform settlers and Land Credit beneficiaries - resettled as a result of dam construction, with an area of up to 1 fiscal module and a gross annual income of up to R$ 14,000.00.
GROUP "B"	- area up to 4 fiscal modules. - at least 30 per cent of family income from farming and non-agricultural activities in the establishment. - gross annual income of up to R$ 4,000.00.
GROUP "A/C"	- Agrarian Reform settlers and beneficiaries of Land Credit who have already taken out credit through Group "A" and have not financed through other groups.
Family farmers with: (former Groups[lt] C", "D" and "E")	- area up to 4 fiscal modules. - at least 70 per cent of family income from farming and non-agricultural activities in the establishment. - preponderant family labour, admitting the occasional hiring of salaried labour, being able to keep up to 2 permanent employees - annual gross income from R$ 4,000.00 to RS 110,000.00
Note 1:	Families with an annual gross income of up to R$110,000.00 and a maximum of two permanent employees are also beneficiaries and qualify as Pronaf family farmers.
Note 2:	For the purposes of the framework, it will be deducted from gross annual income: - 50 per cent: non-integrated poultry farming, sheep and goat farming, dairy farming, fish farming, sericulture, fruit farming and non-integrated pig farming: - 70% : rural tourism, family agro-industries, olive growing and flower growing; - 90% : poultry and pig farming integrated or in partnership with agro-industry.

Source: <http://www.deser.org.br/boletins/Pronaf.pdf>. Accessed on: 25 June 2010.

---

[20] Paraná Agricultural Planning Association (APEPA)
[21] Paraná Agricultural Planning Association (APEPA)

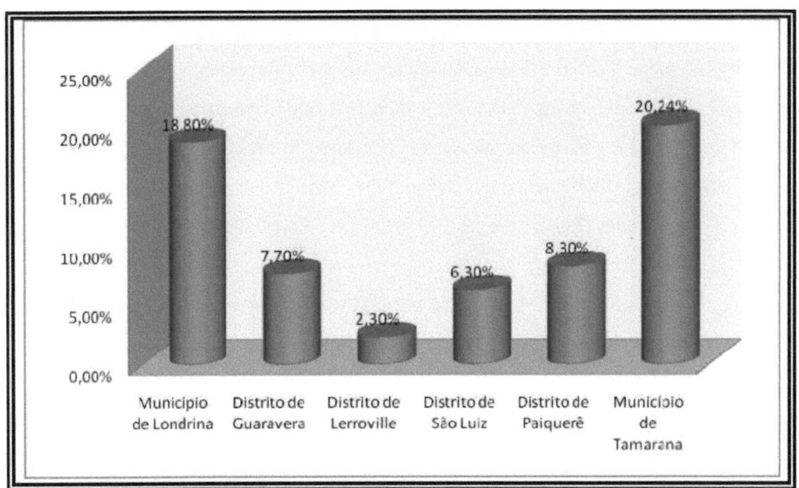

Graph 4 - Origin of Akolá Farm producers
Source: EMATER - Londrina. Dec. 2009.

Graph 4 shows where the families from Fazenda Akolá come from. As we can see, most of them come from areas not far from the farm, especially the municipality of Londrina. However, when we focus on origin, in our interviews carried out on the farm, we found that some families come from other states such as Bahia, Mato Grosso and Mato Grosso do Sul; however, they had already migrated to the state of Paraná before joining the Banco da Terra Programme, and in all the plots we visited, the families were of rural origin. This demonstrates that the Brazilian peasantry "wants to enter the land", as Martins (1984, p. 16) has shown, and that it is a peasantry that, "when expelled, often returns to the land, even if it is a land far from the one it left. Our peasantry is constituted with capitalist expansion, as a product of the contradictions of this expansion".

When asked why they migrated, the answer was unanimous: wage labour or unsuccessful partnerships, meagre earnings, modest welfare conditions and the dream of conquering working land.

The population of Fazenda Akolá is mostly made up of young people. Around 70% of the population is between the ages of 10 and 19. According to data from EMATER (2009), 47 per cent of plot holders believe that their children will continue to work on the farm and only 25 per cent believe that their children will not find a source of livelihood on the farm, but in the city.

However, these figures may not correspond to the reality of the universe surveyed, despite the aspiration of farmers to keep their children on the land, the limitation of land is a determining factor.

We can list the number of children, the small size of the plots and the difficulty of paying off the loan instalments as probable elements that prevent the extended reproduction of the peasantry, which is no different at Fazenda Akolá. As a result, many peasant parents encourage their children to look for other land, either by joining land struggle movements (which is not the case here) or by being included in land financing programmes. This is the case with the Our First Land Programme, which is a special line of credit created by the federal government as part of the National Land Credit Programme, with the aim of financing the purchase of rural properties for landless young people between the ages of 18 and 24 and the children of farmers.

However, despite this family profile, which ends up interfering with the availability of labour, one of the most important crops on Fazenda Akolá is olericulture. According to Filgueira (2000, p. 16):

> The most general and striking characteristic of olericulture is the fact that it is a highly intensive agro-economic activity, in its most varied aspects, in contrast to other extensive agricultural activities, such as grain production. This means that the soil of a plot of land is used continuously, with several crop cycles that develop in sequence.

Field activities take place all year round, while at the same time requiring high investment per hectare farmed, in physical and economic terms. On the other hand, olericulture makes it possible to obtain high physical production and a high income per hectare cultivated and per hectare/year (FILGUEIRA, 2000, p. 16).

However, we must take into account that olive growing is a high-risk activity for rural producers, compared to other agricultural options, due to the greater occurrence of phytosanitary problems (diseases and pests), greater sensitivity to climatic conditions and price instability in commercialisation.

In general, the cycle of olive crops is short, and the same plot can be used throughout the year for three "transplanted tomatoes, or six lettuce crops propagated by seedlings, or even 12 sowings of radish" (FILGUEIRA, 2000, p. 17). Obtaining more than one crop every year means that the physical and economic yields from growing olives are high. The choice to grow olives is also identified by the size of the area occupied, but intensively used, both in space and time.

These possibilities, combined with economic and geographical characteristics, guided the choice of olericulture as a production system at Fazenda Akolá and other Banco da Terra projects studied in the Londrina and Tamarana regions.

At Fazenda Akolá, 99% of the families interviewed produce vegetables for sale. Some of the varieties grown by the farmers on the farm are noteworthy, including yams, cassava, sweet potatoes, aubergines and carrots, as can be seen in Table 4 and the photos below.

Yams, aubergines and sweet potatoes are crops that are favoured in the summer, while carrots and manioc are grown throughout the year. In this way, these farmers are able to produce diversified crops all year round, as long as there are no crop failures. Another important factor for the productivity of these crops is the fertility of the soil. In the photo below (Photo 2), we can see the reddish colour of the structured purple soil.

Photo 2 - Eggplant plantation on one of the plots A, B, C.
Source: EMATER - Londrina. Dec. 2009.

Specifically with regard to the carrots grown on Akolá Farm, we should highlight the favourable morphological conditions for their cultivation. As shown in Photo 3, in the layout of the plots, we can glimpse the planting of green maize located in the lowest part of the farm, where we can see a relief with little slope.

Photo 3 - View of the plantation from plot A.
Source: EMATER - Londrina. Dec. 2009.

We can see in this photo that the topography of Fazenda Akolá is flatter than that of the revegetation where the other Groups studied are located.

Crops such as yam, sweet potato and cassava are heavily produced by these peasants and we can venture to say that there is a certain specialisation in their production, since the majority[22] of the farm's plots have them as their main source of income. As can be seen in Photo 4, the choice to grow yams, as well as sweet potatoes and cassava, was motivated by the high productivity of these crops, which is not only due to the size of the planted area, but also because it is a characteristic of these crops, as well as the considerable ease with which they can be placed on the market, considering both the middleman and commercialisation via the Association or CEASA Londrina. Both for the CEASA "boxistas" and for the North Paraná Association of Horticulturists (APRONOR)

---

[22] With the exception of producer O., who leases his plot to a medium-sized producer near Fazenda Akolá for soya production.

Photo 4 - Cará plantation on one of the plots at Fazenda Akolá
Source: Author's fieldwork. Dec. 2008.

There are producers who deliver their produce to middlemen who sell directly to CEASA in São Paulo, and there are others who gather their produce and sell it via the Akolá Farm Producers' Association. Those who choose this delivery route get a better price for their produce, as Mrtvi (2009, p. 1) pointed out earlier. Carrots, sweet potatoes, aubergines and manioc are also marketed through intermediaries or the produce is sent to CEASA in Londrina by the producers.

In addition to the crops mentioned above, there are other smaller crops on Fazenda Akolá, such as oranges and vegetables like lettuce and cabbage, which are being introduced on some producers' plots. Some plots also produce garlic, chilli peppers, parsley, spring onions, oranges, onions, mint, rosemary and basil for self-consumption.

As well as agricultural production, which the programme is supposed to strengthen once it gives workers access to the land, there are producers on the farm who are engaged in other activities, such as plot T., whose aim is to build a fishing village and a rural restaurant, along with the area used to produce tea and the area already available for a shooting range, thus changing their line of business to rural tourism.

We should point out that some of Akolá's producers are part of the Food Acquisition Programme (PAA), a federal government programme that aims to acquire food from family farmers (who are members of the Family Farming Cooperative

(COAFAS) and meet the PRONAF requirements) for distribution to organisations that serve people in situations of food insecurity.

The programme purchases food, without calls for tenders, at reference prices that cannot be higher or lower than those practised in regional markets, up to a limit of R$3,500.00 per year per family farmer. However, the number of farmers included in this programme is still very small. According to Guilherme Casanova Jr. (2009, s/p.), director of supply at the Municipal Department of Agriculture, there are only 44 associated farmers, who serve 38 entities in the municipality of Londrina, with 25 tonnes of food.

Although the PAA is a public policy instituted in 2003, the programme is moving at a slow pace. According to Casanova JR. (2009, s/p.), "the federal food acquisition programme has been in the process of being implemented since the previous administration. Documentation problems delayed this process, which has now been launched by Mayor Barbosa Neto".

Even though the specialised production of vegetables is notable, at the time of our fieldwork there were two plots on Fazenda Akolá that had started producing tobacco for the Brazilian tobacco company Souza Cruz. At the time of our research, these producers were starting to dry the leaves of their first tobacco crop and were confident that this production would bring them more income than the crops they normally cultivated.

It's worth remembering that these producers have different production conditions, as they have been on the farm since the beginning of the programme and have never been campers or settlers in INCRA's land reform programmes.

These are producers who don't have all the equipment needed for an intensive and diversified activity, so they have to pay for the use of some of the equipment used in production, such as tractors. Due to the distance between the farm and Londrina's Central Supply Station (CEASA) and often the lack and high cost of transporting the harvest, it is not uncommon for producers in the same situation to be forced to hand over their produce to middlemen, which considerably reduces the income from peasant labour.

Although more than 80% of those interviewed own their own vehicles, such as cars and vans, they are often more than fifteen years old and in poor condition; therefore, in the farmers' view, they are not a vehicle for transporting their own produce. On top of this, there is the cost of fuel and tariffs charged for operations inside CEASA, which is often disproportionate to the gross earnings from the activity, forcing them to look for other strategies, including tobacco farming.

On the aforementioned plots, their producers also make use of agricultural credit, the National Programme for Strengthening Family Farming (PRONAF), so that they can fund their crops.

Prior to joining the Banco da Terra programme, these producers were already involved in farming activities, but they didn't own the land, one was a salaried worker on a maize and cotton producing property in the municipality of Londrina. The other also has a tradition in agriculture, but he wasn't an employee. He lived on his father's farm in the municipality of Assaí - PR, where cotton, soya and wheat were produced.

When asked why they chose to grow tobacco, the growers argued that this would be the best alternative for paying off the instalments on the land loan, around R$5,100 a year. The company to which they send their produce, Souza Cruz, would be responsible for the costs related to the initial phase of the crop (seedlings, manure, fertilisers, pesticides in general) and for buying the produce.

One of the farmers told us that he had allocated part of his plot to tobacco production after successive losses of vegetable crops after 2006. This was due to the dispersal of the herbicide 2,4D, used by tenants of Fazenda São João, in the municipality of Pitangueiras, located 86.5 kilometres from the District of São Luiz. Using an aeroplane, these producers sprayed the agrochemical on the crop, which led to its widespread dispersal, causing damage within a radius of 100 km.

According to the leaflet for the product marketed by Nortox S.A.[23] , it is a selective herbicide for use in weed control in wheat, maize, soya, rice, oats, sorghum, sugar cane, coffee and brachiaria pastures. However, vegetables are sensitive to this herbicide, causing the plant to die when in contact with it.

Another characteristic of the product is its high mobility, with great potential for movement in the soil and being able to reach groundwater in particular, making it an extremely strong and dangerous product for the environment, according to the taxological classification laid down in Law No. 7,802 of 11 July 1989.

The drift[24] of the herbicide in 2006 was not the first event of this nature. In 2003, some plots on Fazenda Akolá were affected by the same pesticide, which at the time had been used by producers on the same farm.

After a technical visit, EMATER - Londrina extension workers found

---

[23] 2,4 D Herbicide leaflet.

[24] Drift: this is the deviation of the trajectory of the droplets produced during spraying away from the intended target. The area affected can be another crop, watercourses or any vegetation close to the application site. It is the movement of the product away from the target, which can affect more distant areas. Problems occur when this movement affects a crop that is sensitive to the product applied (SEAB, 2010).

that the herbicide 2,4D had drifted on both occasions, resulting in the loss of crops affected by the product.

The farmers affected filed a complaint with the Paraná Civil Police Department, at the 6th Police District Police Station, by means of a police report. EMATER also carried out an expert loss assessment report, which listed the losses suffered by the farmers. All of this culminated in a lawsuit, which was extinguished after the parties reached an agreement to reimburse the losses.

These situations, to a certain extent, weighed on the decision to diversify the activity and thus integrate into tobacco growing, as well as some facilities offered by the company, such as a 50 per cent subsidy for the purchase of a laptop computer for one of the producers.

Another apparent facility is the cost of the initial phase of tobacco cultivation, which is not a donation from the company, since these costs will be deducted when paying for the production, which must necessarily be delivered to the company, regardless of the price to be paid.

The investment for the production of 3,800 tobacco plants was approximately R$5,000.00. At the time of the fieldwork, the producers were unable to say how much their production would sell for, since the price varies according to the quality of the tobacco leaves. And as already mentioned, they were in the middle of their first harvest.

Information on the average price paid for the production of integrated producers was sought from the company, but to no avail, as it did not provide any data, claiming that the information in the public domain was on its website. There are no indications of average producer prices.

On these plots, the labour force is made up of the whole family, working up to 15 hours a day, according to information collected in the field. These producers don't only grow tobacco, although they do dedicate more than half of the plot to this crop. The plots also produce yams, sweet potatoes, yams, cassava, corn, beetroot, courgettes and vegetables.

With regard to the socio-economic condition of the families at Fazenda Akolá who are integrated into the tobacco production chain, we can say that it contrasts with that of some families who primarily produce vegetables.

In general, the houses of Akolá's families are simple, made of masonry, some mixed (wood and masonry), with 2 or 3 bedrooms, a living room, kitchen and bathroom. As can be seen in the photo below, which represents the most common model of house on the Akolá plots, including that of the producers who started growing tobacco.

Photo 5 - House on one of the plots on the Akolá farm.
Source: EMATER - Londrina. Dec. 2009.

It's worth noting that the houses of farmers who joined the Land Bank Programme with minimal structural conditions are already of a better standard, as can be seen in Photos 6 and 7. The pre-settlement conditions identified were: own vehicle for transporting production, an income to start production without having to resort to land loans which will add to the debts to be paid, tradition in producing the crops proposed in the technical feasibility project. We also can't overlook the differential dedication to crop cultivation among the farmers.

This is easily seen in Photo 6, which shows one of the houses on plots A, B and C[25] . In fact, these are three plots owned by three brothers who, from the start of the group, have worked together to produce and sell their produce. As a result, they have managed to achieve better welfare conditions than farmers who sell their produce on their own. It should be noted that the houses shown were built with the money from the commercialisation of the vegetables produced on the plots.

---

[25] The methodological criterion adopted was not to identify the plots directly.

Photo 6 - House on plots A,B,C.
Source: EMATER - Londrina. Dec. 2009.

Photo 7 - House on plots A,B,C.
Source: EMATER - Londrina. Dec. 2009.

As a rule, the production logic at Fazenda Akolá, as in the other Land Bank Groups, can be understood in the light of the concept of the monopolisation of territory by capital (OLIVEIRA, 2002), which was highlighted in an introductory manner earlier. This generally occurs in sectors that require an intense labour force.

In the logic of the monopolisation of territory by capital, on the one hand there is the peasant who retains control of the land and, on the other hand, capital which

controls the circulation of goods. Thus, production takes place within relations that are not typically capitalist, "in which workers are not deprived of the means of production"

(PAULINO, 2006, p. 103).

We explained earlier that capitalism is unequal and contradictory and therefore creates and recreates non-capitalist labour relations that are necessary for the production of capital. Therefore, it is not only typically capitalist labour relations that can be dominated and reproduced by capital.

According to Oliveira (2003), when capital monopolises the territory without territorialising itself, it creates, recreates and redefines peasant and family production relations. Capital itself creates the conditions for peasants to produce raw materials for capitalist industries, at the same time as these industries enable the consumption of industrial products in the countryside (fertilisers and pesticides in tobacco, feed in poultry and pig farming, for example). Under these conditions, capital subjects the land income produced by peasants to its logic, realising the metamorphosis of land income into capital.

As already mentioned, the peasant class controls both the means of production and the labour force. It is this specificity that differentiates it from other classes and makes the peasantry part of the market exchange system through the sale of its production and not its labour power, like wage earners.

In this regard, Paulino (2006, p. 108) observes: "[...] what peasants sell under capitalism is the product in which the family's labour is contained, an essential distinction from other workers, who have only the commodity of labour power to sell".

In wage labour relations, there is the formal and real subjection of labour to capital, through the separation between the worker and the means of production. However, to the extent that the peasant maintains control over the means of production, especially the land, and produces on it using his labour and that of his family, and at the same time has a close relationship of dependence on capital, the income from the land is subject to capital.

MARTINS (1995) apud Paulino (2006, p. 110,) defines:

> [...] the notion of the formal subjection of labour to capital is originally related to the expropriation of workers [...]. This subjection would not represent any change in the labour process. It would continue to be carried out exactly as it was in domestic craft production. Only now, the craftsman, transformed into a wage labourer, no longer works for himself, but for the capitalist [...]. The next step is for capital to seize not only the result of labour, but also the way of working [...] in the real subjection of labour to capital, knowledge is restricted to a small aspect of production [...]. To the extent that the producer retains ownership of the land and works it without the use of wage labour, using only his own labour and that of his family, at the same time as his dependence on capital grows [...] we are faced with the subjection of land income to capital (emphasis added).

The monopolisation of territory by capital is used as an artifice by industrial capital, which needs raw materials from agricultural production, but has no intention of territorialising itself, i.e. immobilising a sum of money in the purchase of land, becoming a single agent of capital - industrial capital and the landowner. By monopolising the territory without territorialisation, the capitalist does not own the land; however, he does create the conditions for subjecting peasant agriculture to land rent where it apparently does not exist (PAULINO; ALMEIDA, 2010, p. 45).

> The monopolisation of the territory (monopoly on circulation) expresses precisely this process of organisation and use of the territory by industrial capital. In this way, a certain part of this territory is being occupied/exploited for the production of a certain agricultural product by various producers - mostly small and medium-sized - who have, in a way, lost their economic autonomy and almost always become dependent on the processing industries, which are the ones that make production viable and not the direct producers, as a rule (THOMAZ JR. 1988, apud PAULINO; ALMEIDA, 2010, p. 45).

This artifice is generally used in tobacco, chicken and silkworm production; in other words, by companies interested in raw materials from agricultural production.

> It is essential for the companies that the tobacco grower produces his livelihood. When he becomes a tobacco farmer, contradictorily, he has to continue producing his own food. The fact that he is a tobacco farmer does not mean that he is no longer a peasant. He is paid little. They are paid a low price for the product of their labour, i.e. they are not remunerated at the levels that the price of the product on the market would allow, which demonstrates the process of transferring income from the product to the companies (ETGES, 1991, apud PAULINO; ALMEIDA, 2010, p. 45).

In this group we had the opportunity to meet families living side by side in conditions of re-creation as peasant families, families who do not dedicate themselves to agriculture as their main source of income, and also families who find it extremely difficult to pay off the debt contracted when they bought the land.

These are, therefore, the problems of a programme that has focused solely and exclusively on financing land, without allocating any funding to productive projects, thus making it impossible, on several occasions, for the families involved to recreate themselves.

3.1.2 King of Lettuce

The King of Lettuce Group is, among the Groups studied, the one with the most privileged location, from the point of view of production outlets and access to urban goods and services. The name itself is indicative of the production specialisation within it: the production of lettuce among other vegetables. Located in the district of Warta and three kilometres from the urban area of the municipality of Londrina, the group is between two important state highways linking municipalities in the north of Paraná, the PR-445 and PR-545, as can be seen in Figure 5. It also shows that the plots are located in a region that is surrounded by properties that produce soya (in summer) and wheat combined with maize (in winter), crops on which pesticides are usually applied, some of which are incompatible with some vegetables.

Figure 5 - Floor plan of the King of Lettuce Group.
Source: Google Earth, accessed on 25 October 2010.

Figure 5 shows the proximity to the Warta district and the PR-445 and PR-545 state motorways. It can also be seen that the group is surrounded by soya producing properties (at the time the image was taken).

The property on which the Group is located has an area of 21.78 hectares, with each plot having an average area of 3.63 hectares. The site has flat topography, good fertility, with a predominance of structured purple soil. It has a spring that gave rise to a dam that is used to irrigate part of the plantation, as shown in Photo 8.

Photo 8 - Water pump for irrigating the plantations
Source: Fieldwork. Feb. 2010.

With regard to the Permanent Preservation Area (APP) of the King of Lettuce Group, it is located precisely in the water catchment area, as illustrated in Photo 9.

Photo 9 - View of the right bank of the reservoir and, in the background, the permanent preservation area.

Source: Fieldwork. Feb. 2010.

Negotiations to set up the Group began in 2000, but the land credit was only granted in 2001. The value of

The total funding for the project was R$237,000.00, of which R$39,500.00 was the individual value of the land debt.

The entire process of organisation, registration and negotiation was carried out directly with Banco do Brasil. This demonstrates an advantage over the other cases studied, and even over groups in other regions of the country, because, as already detailed in Chapter 2, it is rare for an association to achieve autonomy in land negotiations and commercialisation.

The land credit in this Group, unlike some of the cases studied, was carried out individually, so there is no joint guarantee. Another unique feature of this Group is that all the members are related, which favoured the negotiations for implementation, as well as the continuity of the Group until the outcome detailed below.

The fact that the "beneficiaries" of this group were already traditional vegetable producers, even working in this area as tenants, meant that they didn't need to resort to other lines of agricultural credit such as PRONAF, so that their only debt was the purchase of the land. This was because they already had the necessary tools for production, as well as having small plots of land in the Londrina region, which were sold so that the money could be invested in the plots. This provided a certain level of well-being, as well as sufficient monetary income to pay for the financing of the plots.

The levels of capitalisation capable of ensuring the families' considerable well-being were only achieved thanks to their inclusion in the Banco da Terra project, since owning their own land, despite the burden of buying it, allowed them to transcend the condition of tenants. However, there was already a difference here, because prior to joining the project, they had minimal structural conditions for their recreation, namely: houses, barns, running water, vans, work tools and irrigation equipment.

An on-site investigation revealed that the Group is equipped with all the agricultural implements needed for vegetable production, such as a planter, tractor, subsoiler, enchanter[26] , irrigation system (for part of the production), greenhouses, sheds for storing the produce and a lorry for transporting the produce, as shown in Photo 10.

---

[26] Artefact used for planting seedlings, similar to the nunchucks used for planting cereals.

Photo10 - Greenhouse, lorry and storage shed

Source: Fieldwork. Feb. 2010.

Quadrant A shows the greenhouse, quadrant B shows parsley seedlings. The lorry in quadrant C is used to transport the produce. Finally, quadrant D shows the shed where the produce is stored.

It's worth noting that when the photo above was taken, the greenhouses were almost empty, as all the seedlings had been planted.

The King of Lettuce was made up of six families originally from São Carlos do Ivaí, but who had lived in the Londrina region for over twenty years. The families already had a tradition in agriculture, more precisely in the production of vegetables, both as smallholders and as tenants and/or wage labourers in the field.

Due to the significant changes that have taken place in this Group, it is appropriate to divide our analyses into two parts.

At the beginning of our research on the Group, in 2008, two families were no longer part of it, as they sold it after paying off the debt from the purchase of the land from Banco do Brasil. One plot was sold to soya producers and the other was bought by one of the beneficiaries of the same Group. In Figure 6, dated 15 May 2006, we can see the first spatial configuration of the King of Lettuce, which was modified after some families left the Group, as we'll discuss later.

Figure 6 - Configuration of the King of Lettuce plots (on 15 May 2006)
Source: Google Earth, accessed on 04 Apr. 2011.

Plot 1 stands out in the figure, represented by A. B and C represent the communal vegetable production area and the area of the houses on plots 3 and 6. The dam and the permanent preservation area are represented by D. E and E1 represent the rest area and the communal coffee production area respectively. The area of the houses on plots 2, 4 and 5 is represented by F.

In this image we can see that the production areas were communal. Even though the areas of the plots were separate, all the commercial and self-consumption crops were produced in the same area by everyone in the group.

At that time, the plots combined production for self-consumption and the market. On the plots, various crops were produced communally, such as corn, rice, beans, coffee and varieties of pulses, and chickens and pigs were raised for family consumption. According to reports from members of the group, they only bought personal hygiene and household cleaning products at the supermarket.

Market production was specialised in olericulture, producing lettuce, cabbage, kale, parsley, spring onions, broccoli, rocket and chard for hypermarkets, small markets and greengrocers in the municipality of Londrina. In this way, we had plots of land where peasant ownership of the land and agriculture aimed at satisfying the needs of the individuals who interact in the production unit were combined with commercial agriculture as a way of providing the monetary increases needed to buy the goods indispensable for the family's reproduction.

We therefore found that there was a peasant logic, in which the family's reproduction takes place in two dimensions: economic production and social reproduction; in this logic the enterprise, understood as a production unit, is not merely economic, as it is in the capitalist logic of reproduction.

At that time, patriarchs lived on the plots and managed to enable their children (beneficiaries in the same group) to continue the peasant logic, in other words, the extended reproduction of the peasantry.

Although there were permanent salaried workers employed on the plots at that time, the predominant labour force was that of the family.

In 2010, when we returned to the field for further investigations, we realised that two other families had left the area. One was that of one of the patriarchs of the original Group who, after paying off the land purchase debt in full, sold it to one of his sons, a beneficiary in the same Group, and returned to his home town, São Carlos do Ivaí, where he acquired a small property where he also produces vegetables that are sold at the town's open markets. According to information collected in the field, he opted to do this because he could get a better price for his produce, since the product is marketed directly to the consumer and no longer to retailers, as is the case with those who remain at Rei do Alface.

The family of another patriarch of the original group, made up of two children and his wife, also left the King of Lettuce after the debt was paid off. Likewise, this plot was sold to the same beneficiary who had acquired the land in the negotiations mentioned above. Although it was not possible to collect precise information on the fate of this family, it seems that family breakdowns prevented them from continuing to recreate themselves as peasants.

Currently, there are only two families from the initial group at Rei do Alface. The following photo (Photo 11) shows the home of one of the families, which expresses the different income conditions enjoyed by the King of Lettuce producers compared to the other Land Bank beneficiaries studied.

Photo 11 - House on plot 1, located a few metres from the PR545.

Source: Fieldwork. Feb. 2010

It's worth noting that, despite the capitalisation process achieved by expanding the olive-growing activity in the terms presented, the family of the main manager of the Rei do Alface unit still lives in the same house, which was already occupied when they were tenant farmers in the area, as shown in the photo below (Photo 12).

Photo 12 - House on plot 3
Source: Fieldwork. Feb. 2010.

The house in question is an old dwelling, already used by the farmers when they were still tenants.

This is evidence of the detachment from the signs of modernity and ostentation that some peasants retain, which is why class logic cannot be taken on the basis of purely economic criteria.

As for the composition of the two families, one of them is made up of four people who actually live on the land, a father, mother and two children. The profile can be considered young, as the ages range from 12 to 38. The level of schooling is considered high compared to other cases studied, since one of the owners has completed primary school and the other secondary school, and the two school-age children attend school a few metres from where they live, in the Warta district. The other family is made up of a father, mother and a teenage son. The parents have finished secondary school and their son is regularly enrolled at the school near their home. In the photo below (Photo 13), you can see the children of both families.

Photo 13 - Children of the King of Lettuce producers.
Source: Fieldwork. Feb. 2010.

There is no doubt that the composition of the families is a factor that forces them to hire workers, since olive growing is a labour-intensive activity and the children are not yet old enough to work. Nevertheless, the increase in salaried labour has profoundly transformed the Group's previous structure.

Previously, almost everything needed for food was produced on the plots. Today, this is no longer the case, with the exception of a few pig and chicken farms and a small area of coffee and maize. However, we can't rely on this to categorically point to a breakdown in the peasant labour that existed there, because we have to consider that the production that was destined for self-consumption (rice, beans, maize, coffee and some legumes) at that time was opportune, since there were more families engaged in this cultivation and the cost of this production was lower than the purchase price of the same industrialised products. This logic had already been pointed out by Chayanov, who indicated strategies for substituting self-consumption production whenever the purchase could compensate for the commitment to production, without proportional monetary losses.

With the departure of the four families that made up the original group, the area that had previously been set aside for planting those crops was turned over to growing vegetables. This happened for two reasons: firstly, because there was a reduction in the number of members, which led to a shortfall in the amount of labour, and also because the families that remained preferred to extend their commercial production and no longer produce certain products such as rice and beans, as they would have a higher production cost than those available on the market.

To summarise, on one of the plots three people work regularly, the owner and two salaried workers. On the other, five people work, including Mr D, his wife and three salaried workers. Both plots have permanent salaried workers, but only one has a labour registration card. When asked why he didn't register the other employees, Mr D. told us that it's difficult to register all the workers because the charges are high.The changes mentioned can be identified in the spatial configuration of the Group today, as shown in Figure 7.

Figure 7 - Current configuration of the King of Lettuce plots

Source: Google Earth. Accessed on: 25 October 2010.

Plot 1, represented by A, stands out in the figure. B and C represent the communal vegetable production area and plot 3, the APP and area of

E and F represent the communal coffee production area and the area of the houses on plots 2 and 4 (which are unoccupied), respectively. Finally, the blue polygon highlights the area of the plot sold to non-beneficiaries of the Group, where soya is currently grown.

We can see that after the patriarchs left, the Group changed its working relationships, because if we looked at Rei do Alface, we could find a peasant economic unit there, not only because of the commercial production combined with self-consumption production, but also because of the strong presence of family labour. Today, the group has lost the main characteristic of a peasant economic unit, family labour as the main agent of commercial production.

Referentially, classes in the capitalist world are defined according to the origin of their income. There are the capitalists, who appropriate the surplus value of others in order to make a profit; the proletarians, who, deprived of the means of production, have only their labour power as a commodity, which is traded in exchange for wages and, finally, the landowners, who earn the rent from the land. Obviously, this is a parameter, as Shanin (2008) rightly pointed out

If we were to maintain a linear reading of society, we would consider that because wage labour supplants family labour, determining that the source of these beneficiaries' income comes from the exploitation of surplus value, we would characterise them as capitalists. However, we won't dare to classify these individuals, since they are the same peasants as yesterday; we will therefore analyse the Transition Group.

Ploeg (2008) makes some useful points in this regard by invoking the concept of peasantry to understand production units, given the internal dynamics that are deeply determined by family labour and its interference in the economic strategies undertaken by peasants.

This is not a new assertion, not least because Chayanov (1974) chose the demographic dynamics of the family as the founding element of the theory of the balance between labour and consumption, in view of the profound changes it imposes on productive units according to the stage of constitution and the age group of each of its members.

Without disregarding the context on which the author was based, as well as the specificities of the production units forged by the Banco da Terra Programme, this variable cannot be ignored in the strategies adopted, in which the complementary nature of the hired workforce cannot be gauged in quantitative terms, given the dynamics observed in the production units in question.

So we can't say that the current labour situation in the Group will be perpetual, because the monetary situation and the need to provide for the family's demands are not stable elements. At some point, these producers may lose their main buyer, who currently accounts for more than 80% of their production, and have to restrict production, which means going back to using family labour exclusively, or they may even lose their means of production and return to being tenants or salaried workers. Or they may be able to capitalise in such a way that they can acquire more land and move into large-scale production based on wage labour.

The fate of these producers holds infinite possibilities, and this doesn't depend on them alone. In our field visits to the Group, we were able to see that they do not plan, at least explicitly, in the short term, to accumulate land and change production. Apparently, this transitional situation has occurred because these producers have managed to reach such a level of capitalisation that they don't need to burden the whole family in order to carry out production.

When asked if they had ever thought about changing the cultures they produced, or even changing professions, they reported that they had,

> Look, I spent four years off the land, I went to Mato Grosso do Sul, when I was still renting here with my father, to work as a cane-pulling lorry, I even made a bit of money, but it didn't pay off. Then I came back and it worked out that I was going to do the Banco [da Terra] deal, and we're still here today. But, honestly, when there's a crisis, I think about dropping everything and going to São Carlos do Ivaí [PR] to work with lorries again. I've even thought about it today[27] , because, look, there's almost nothing there [for planting], but then I think about the children, about starting all over again, then I try a bit harder and then it gets a bit better and we get on with it. (D., Grupo Rei do Alface, January, 2010)

In affirming that the fundamental characteristic of the peasant class is family labour, we are guided by the study of Chayanov (1974, p. 98) in which the author states that, when analysing a peasant economic unit, one finds above all that one of its organisational elements, the workforce, is fixed, since it is present in the composition of the family. Thus, "[...] in the peasant domestic exploitation enterprise, the family's labour force is something given and the productive elements of the unit are fixed according to it". In this way, the author states that the size of the family determines the intensity of

---

[27] In the first months of 2010, when the fieldwork was carried out, there was a high level of rainfall, which led to many losses in the plantations.

labour (self-exploitation) and the composition of all its elements.

> It cannot be increased or decreased at will and, as it is subject to the need for a correct combination of factors, we must naturally place the other factors of production in an optimum relationship with this fixed element, which places the total volume of our activity within very narrow limits. (CHAYANOV, 1974, p. 98)

Despite the changes indicated in terms of labour relations, production continues to specialise in vegetables, as can be seen in Photo 11 below, which shows the planting of lettuce, one of the flagships of the production unit in question.

Photo 14 - Lettuce bed. **Source:** Fieldwork. Feb. (2010)

As for the capitalisation of these producers, this wasn't due to an increase in the selling price of their products, because that decision isn't made directly by the producer, but by the buyer of the produce. What has happened is an increase in the placement of products on the market. Both to supermarkets and to local sales.

At Rei do Alface, produce is sold in three different ways: to supermarkets in Londrina, which is the Group's main sales channel, responsible for 80% of the products' destination; to CEASA; and to people who come to the Group. The value of the sale is low, both for supermarkets and CEASA, as shown in Table 7. We can also see in this table that when the producer has the autonomy to commercialise his products, as in local sales, the prices are fairer, because it is the farmer who dictates the price. However, this means of commercialisation only accounts for 10% of the total marketed, but the monetary return is higher, as the value is up to four times greater.

Table 7 - Production and Commercialisation Prices for the King of Lettuce

Product	Production price per unit (approximate value per unit)	Selling price (Supermarkets)	Selling price (CEASA - Londrina)	Selling price (on site)
Lettuce	R$ 0,35	R$ 0,50	R$0,40-0,70	R$ 1,00-1,50
Chard	R$ 0,40	R$ 0,80	-	R$ 1,50-2,00
Broccoli	R$ 0,70	R$ 1,00	-	R$ 3,00
Kale	R$ 0,30	R$ 0,50	R$0 ,40	R$ 1,00
Cauliflower	R$ 0,40	R$ 0,80	-	R$ 1,00-1,50
Chives and parsley	R$ 0,20	R$ 0,45	-	R$ 0,70
Cabbage	R$ 0,40	R$ 0,70	R$0 ,80	R$ 2,00
Arugula	R$ 0,30	R$ 0,80	R$1 ,00	R$ 1,50

Source: King of Lettuce producers, interview conducted by the author in February 2010.

When selling to CEASA and supermarkets, the price is set by the buyer, and at this point the producer has to submit to the trade, as these prices don't tend to vary much from one supermarket to another. When we asked the producers why they continue to sell to supermarkets, we learnt that:

> [...] We've been selling to Supermuffato for a long time, ever since we were tenants here. My father knew the people there and so we started taking our produce there. At first it was difficult because we had to take the products there very early in the morning, put them on the shelves, everything just right, it was very tiring, because it wasn't just one shop, it was three. Now, since there's a distribution centre, we take it there and the staff tidy it up. So even though the price is low, sometimes R$0.17 per unit [sales value minus production value], it pays off, because I know they'll always buy from me. If I go to sell elsewhere, I could be exchanged for something else, which is why it's better to leave it the way it is. We sell almost everything there, sell a little to CEASA, when there's enough, and sell it here at home to the people who come here. Actually, I'd like to sell everything here, but unfortunately demand is still low [...] (D. Rei do Alface, Feb. 2010)

We can see in this statement that even though the producer earns little for his product, he continues to sell it in the same place for security reasons. The fear of having nowhere to sell their produce is latent in all the producers in the Groups studied, as they recognise their duty to pay off the land debt.

The King of Lettuce is the exception to the Programme's rule of failure, because only in this group can we see that they manage to produce in order to pay off their debt and provide for their family's survival. However, these producers, like all the others studied, are subjected to capital in all areas. Whether it's in the purchase of inputs for production (industrial capital), in marketing due to low prices (commercial capital) or because they are part of a land financing programme via the Bank of Brazil and the World Bank (financial capital), it's clear that peasant agriculture can only be analysed in the light of the contradictions that sometimes allow it to recreate itself, in which case the Bank of the Earth policy applies, and sometimes impose its disappearance; either through impoverishment and reduction to proletarian status, or through enrichment that elevates it to bourgeois status. Hence the need to break away from a linear reading of reality and pay attention to the processes that characterise the countryside, which are in permanent movement.

3.1.3 Alto Alegre Group

Located in the district of Lerroville, also in the municipality of Londrina, the Alto Alegre group is made up of 47 families. The total area is 196.92 hectares, comprising 3.86 hectares per plot.

In this group, the families are not related and come mainly from around the municipalities of Londrina and Tamarana, with some coming from rural villages[28] in this region. The families prioritise the production of coffee and vegetables, combining production for sale and self-consumption.

We should also point out that, unlike the other projects studied, this group did not have the Paraná Institute for Technical Assistance and Rural Extension (EMATER) as a mediating agent in the constitution process. The technical and production project was carried out by the company Planejamento Agropecuário LTDA (PLANATEC) which, after the projects were realised, no longer offered technical assistance visits or any help for them, a fact revealed in the interviews with the families. In an interview with the MDA in August 2010, we were told that the company, two years after setting up the group, had still not sent the definitive maps and data from the topographical survey of the area. From the outset, this reveals irregularities in the way the project was conducted, because without this data, the demarcation and occupation of the plots could not have been authorised. Thus, the plots were demarcated according to the preliminary topographical surveys.

We should also point out that the company didn't provide any data on

---

[28] The Rural Villages Programme was set up in 1994 by the Paraná state government and its main objective, according to Reis (1997, p.4), was "[...] to help rural working families who are 'bóias-frias', settling them in their own environment, guaranteeing them housing and land so that they can get out of the miserable condition in which they live". However, studies such as Paulino's (2006) show that these objectives have not been achieved.

Alto Alegre; all the data that will be presented was collected from the MDA and on site. We tried several times to get in touch with the two technicians responsible for the Group, but we never managed to speak to them. We also tried the company, but to no avail.

We found it very difficult to interview some of the families because, at the time of the field surveys, in February 2010, there was heavy rainfall in the region studied. Due to the very steep terrain, there was a landslide forming a ravine located on the road that connects the plots, which made it impossible for us to visit these plots. However, we did have the opportunity to talk to some of the residents of these plots at the event held at Fazenda Akolá in December 2009.

This group is made up entirely of middle-aged people (between 40 and 50 years old). However, there is one plot where the owners are very young, a couple aged 24. The level of education is low, considering that more than 50 per cent of this population has not completed primary school and only 20 per cent has completed secondary school. School-age children regularly attend public schools in Lerroville or Tamarana.

As previously mentioned, the production of these plots follows the technical or economic viability project carried out by PLANATEC, in which the beneficiaries were supposed to produce Arabica coffee. As we've already mentioned, the economic viability project is conceived with the sole aim of adopting a crop that can be produced in small areas and that can generate enough income to pay off the debt. No consideration is given to whether the farmers have experience growing this crop or even if they have the necessary tools for this production.

Considering that coffee productivity is biannual, meaning that one year there is a good yield and the next year production drops considerably - added to this, we should point out that it takes an average of three years for a coffee tree to start bearing fruit - economic viability is questionable, which is why the other projects studied favoured the cultivation of vegetables, which have a faster monetary return.

As the first instalment of the financing will be due by the end of 2011, the year in which the coffee plantations began to bear their first fruit, it appears that default will be a necessary consequence of this phase of the project, if not the others. The photo below illustrates the stage of this crop in the Alto Alegre Group.

Photo 15 - Coffee plantation still in formation on one of the plots
Source: fieldwork. May 2010.

Faced with the lack of income until then, in view of the technical indication for this crop, the alternative found by some producers was accessory labour, as tractor drivers or as salaried workers on farms in the Londrina and Tamarana region, since the production of a few crops for self-consumption alone does not allow these families to survive. The crops grown for self-consumption are manioc, maize, beans and vegetables.

When we carried out our field surveys, there was only one plot that didn't follow the technical project and was commercially producing vegetables such as cabbage, lettuce, chicory, rocket, parsley and spring onions. This is a very young couple, only 24 years old. It's worth noting that this couple is trying to reproduce in the project the possibilities of peasant recreation experienced by their parents in the Moreiras neighbourhood, located in the rural area of Tamarana, which is a community of small rural landowners who are dedicated to growing vegetables. It is an eminently peasant neighbourhood, with small properties, most of which do not exceed 3 hectares, which is why olive growing has become an important part of their lives.

as a viable recreation strategy. It is the existence of dozens of such units that has made it possible to transport the produce to Ceasa in Londrina, and thus to continue the activity. Because they have experience in growing vegetables, the couple chose not to pursue the economic viability project. They told us that at no point did they think about planting coffee because they had no knowledge of the crop. What's more, they consider the choice to be the right one because

> [...] while those of us who are growing vegetables have managed to save a bit of money to pay the bank, the people who are growing coffee are having a hard time, because the plants are still small. The moment they start producing, that's when they'll have to pay the bank, and then they won't be able to cope. When we arrived here, people didn't believe we were going to grow vegetables and said we had to grow coffee because it would be better. As we were new, people thought we were going to break our backs. But thank God that didn't happen. We're producing, we've managed to finance a small truck so we can take our produce to CEASA, and we've even rented part of our neighbour's plot upstairs to plant more.

We can see in this statement that the fact of not following the economic viability project is not necessarily related to the failure of the plot, as is pointed out by some technicians and the MDA itself. We can say that in this case, it is precisely the discarding of the economic viability project that is the reason for the plot's success.

As the couple mentioned, they sell their produce at CEASA. Along with commercial produce, the plot also produces maize, cassava and beans, which are used for self-consumption.

The couple work on this plot, along with two other salaried labourers, who are not registered. The couple told us that the two of them alone wouldn't be able to produce enough to earn a good income, because olive growing is very intensive, which means they have to hire helpers.

This group is the one that best represents the mismatches of the Land Credit Programmes in Brazil, both because they were planned and carried out by a private company, and because they had a productive or economically viable project that was conceived based solely on the income that growing coffee could generate, but did not take into account how these families would survive and pay off the land debt until the sale of the product generated income. As a result, the families who were unable to find other sources of income at the time of the fieldwork were in the extreme situation of wanting to abandon the plot, thus demonstrating the inefficiency of the programme.

3.1.4 Groups Renascer I, Renascer II, Renascer III

Even with the presence of peasant resistance in encampments, INCRA ceased negotiating areas for settlements in 1999 (TSUKAMOTO, ASSARI, 2005), thus favouring negotiations with Land Credit Programmes aimed at financing properties. It is in this context that we can understand the installation of Renascer Groups I, II and III.

Photo 16 - Camp organised by the MST.
Source: fieldwork. June 2010.

It's worth remembering, however, that the demands for land were not met by the programme and evidence of this are the encampments that still exist in the region.

The camp in question is located at the entrance to the Serraria settlement, next to the Esperança Group. This camp is home to 200 families hoping to be settled in Londrina or Tamarana.

It should be pointed out that the settlements and Land Bank Groups are generally located in areas with predominantly eutrophic lithic A soils, which are poorly developed, have low fertility, are shallow, occur in places with steep slopes, are often littered with gravel and rock fragments and are highly susceptible to erosion. It is worth returning to Figure 3, which better illustrates the pedological conditions of the areas in question.

Properties located in areas that are difficult to access, with low soil productivity and rugged terrain, were negotiated by INCRA or Banco da Terra directly with the owners, according to the testimony of some beneficiaries of the Groups studied, who were workers on the negotiated properties.

The characteristics of the predominant soil in Tamarana and the subjection of producers to the market are one of the factors that make it impossible for the majority of beneficiaries of Land Bank Groups to pay off the land financing, production costs and PRONAF debt[29] .

Many families are unable to put their plots to work, either for lack of labour or lack of money to buy what they need to produce commercially. So, in order not to lose the land that many people have been waiting for, many people go in search of work, either as wage labour in the countryside or in the city. We see this as a peasant survival strategy. In this way, the search for temporary work outside the plot does not point to a disintegration of the peasantry, but rather a way for them to remain peasants and maintain their only means of production, the land they work.

Paulino (2006, p. 324) states that [...] the inexorable thesis of proletarianisation is lost in the face of the vast evidence that this dynamic even involves a strengthening of the class, brought about by the acquisition of land.

The criterion for measuring the proletarianisation of the peasantry cannot be based on the level of well-being in terms of the volume of material goods, but on the way they organise themselves internally. Peasants cease to be peasants when they losing control over the means of production or incorporating the capitalist logic expressed in the measurement of profit through the exploitation of labour (SHANIN, 2005)

For Chayanov (1974), the fact that peasants look for ancillary work is not necessarily a destructuring of the class, but rather a strategy for maintaining peasant status. The author also states that the displacement of labour from agriculture to non-agricultural activities can occur when these activities offer higher earnings than agricultural ones.

> [...] in many situations it is not a lack of means of production that leads to earnings from crafts and trade, but a more favourable market situation for this type of work, in the sense of the remuneration that peasant labour provides compared to agriculture. (CHAYANOV, 1974, p. 118)

Starting from this common scenario in Tamarana, where accessory work is imperative, we will analyse the groups of Renascer I, Renascer II, Renascer II, Brasil and Esperança.

The Renascer I group is located in the municipality of Tamarana, and is made up of 22 families who are allocated a total area of 100.9 hectares, comprising an area of 4.2 hectares per plot. The purchase price of the property was R$599,999.83, and the total value of the financing was R$758,387.33. The value of the financing was

---

[29] We generalise PRONAF debt, because 95% of the borrowers in the Groups studied have acquired this line of financing, both for costing and for infrastructure and credit.

calculated on the basis of the purchase price of the land and the expenses[30] and investments to realise the Group, namely: basic infrastructure (construction of 24 masonry homes with 48m^2 each, with a total value of R$ 120.000.00), a water supply system for human and animal consumption (R$7,600.00); setting up an internal electrification network (R$7,000.00); opening, rehabilitating or building three internal access roads (total value R$6,187.50); notary costs (R$14,000.00); topographical costs, demarcation of plots (R$3,600.00).

The process of registering and submitting documentation for the loan took place in mid-2003. However, the plots were only occupied in March 2004. This was because the financing contract was collective (joint and several guarantee) and if any of the beneficiaries had outstanding debts with the Federal Revenue Service, the Centralisation of Bank Services, or any other authorities, they would have to be paid.

S/A (SERASA), the Credit Protection Service (SPC), among others, the financing proposal sent to Banco da Brasil went back to the union. As a result, until the situation was resolved, the proposal was put on hold, since, according to the dictates of the Programme, there could be no substitution of beneficiaries.

This group originated from the fragmentation of a group of 59 people interested in acquiring land via the Land Credit Programmes to finance rural properties. Demand for the Tamarana Rural Workers' Union was high, both from workers wishing to apply for land financing and those interested in selling their land in Londrina and Tamarana. This demand on the part of landowners is justified, since the price paid in cash by the Banco da Terra Programme was 10% of the market value (NABARRO, 2007).

It should be noted that none of the groups formed in Tamarana had autonomy in choosing where they would live. This was a precept of Banco da Terra.

The group of 59 people gave rise to the Renascer Groups (I, II and III), with 24, 18 and 17 families respectively. However, due to environmental legislation, two plots in Renascer I had to be converted into a legal reserve, as the area previously earmarked was smaller than required by law. As a result, two families were already bound by all the documentation in the contracts sent to Banco do Brasil, including the collective deed for acquiring the land. Thus, two families took on a debt of R$31,600 each, but for a plot that they will never occupy.

Most of the residents of this group come from the municipalities of

---

[30] All the figures were taken from the acquisition and financing contract for Grupo Renascer I, available from EMATER - Tamarana.

Londrina and Tamarana, and those who come from other states have lived in the region for between 10 and 20 years. 83% of the population of Renascer I are of rural origin, and before joining the Programme, they were tenants, worked in the partnership system, salaried workers or campers.The age range of the Group's residents (Graph 5), we can consider that it is not a very young population, since 56% are between 41 and 60 years old.

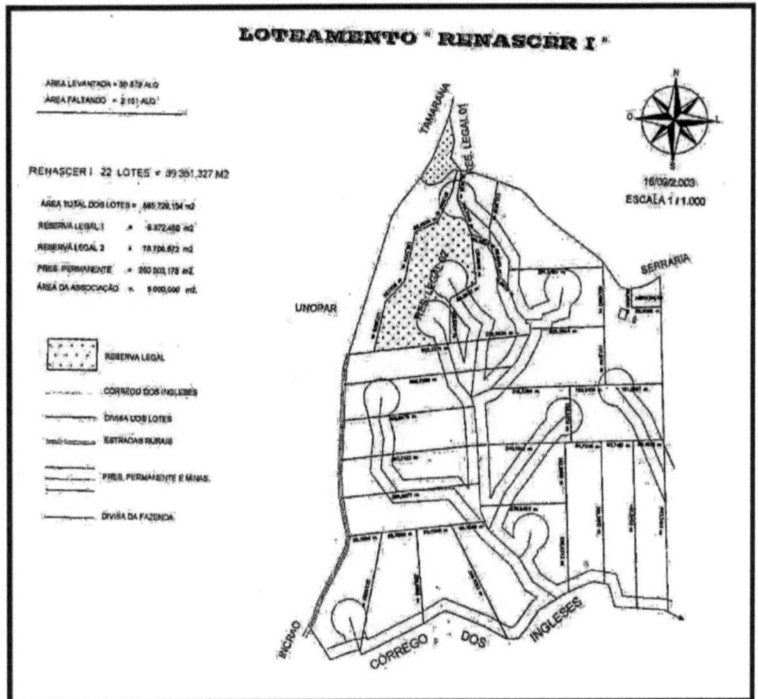

Figura 8   - Renascer Group I plant
Source: Emater (2010)

The level of schooling is considered to be low, with 51% of the population not having completed primary school, which ranges from grades 1ª to 4ª. Illiteracy is low and related to older people. Eight per cent and four per cent respectively represent the percentage of the population that has completed primary and secondary school, this type of education is more related to the younger population.

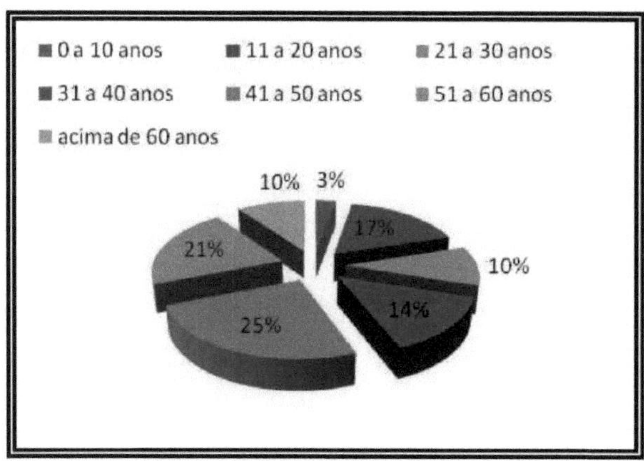

Graph 5 - Age group - Renascer I group
Source: field data collection, June 2010.

Agricultural production on the plots is geared towards growing vegetables, milk, coffee and also maize, cassava and beans, which are used for the families' own consumption.

The production of vegetables is favoured when it comes to commercialisation, as their cycle is generally short, thus providing volume and product diversification. Commercial products include green beans, cucumbers, cabbage, cauliflower and chillies. All the plots visited produce maize (used when green for human consumption and after drying as animal feed) and raise chickens.

In most of the plots, the predominant activity is horticulture. However, there are also members who dedicate themselves to dairy farming in the co-operative system. They sell their produce for R$0.45 a litre, and are subject to this low price because the Londrina Agricultural Cooperative - Cativa collects the produce from the plot. However, the costs of maintaining the herd are constantly increasing. However, the farmers have to submit to the price imposed by the co-operative. Just as is the case with dairy farming, the same is true of olive growing, even if the agent who takes the farmers' income is the middleman or CEASA. The photo below (Photo 17) shows diversity as a peasant self-protection strategy.

Photo 17 - Vegetable production.
Source: fieldwork, June 2010.

The differentiation expressed by the photo is a hallmark of the Group's productive plots. When asked what was produced on the plot, Mrs D. and Mr E. said,

> we plant and harvest everything that doesn't cost a lot, there's corn and manioc for the house, chilli peppers, jiló, green beans that we sell. If we plant something that makes money, like coffee, we won't. Because if we don't, we'll have even more trouble paying for the land, and if we get more PRONAF, we won't be able to afford it (sic).

Reflecting on what the couple said, we realised how difficult it was to grow some crops that could generate a higher income for the residents. Despite all the difficulties, this couple and all the people interviewed were satisfied with their plot. However, a latent concern among everyone is paying off the debt they acquired with the dream of working land.Around 80 per cent of the producers in Renascer I sell their produce to middlemen. A few, who own a truck, take their produce to CEASA in Londrina, a distance of around 80 kilometres. Depending on the Group's location, such as in Brazil, this distance exceeds 100 kilometres. This shows that it is often more advantageous, contradictorily, for producers to be subject to the low prices paid by middlemen. On average, produce is sold at R$10.00 a box, whether it's jiló, green beans or cassava. Leafy greens, such as lettuce, cabbage and cauliflower, are sold at an even lower price because they are more perishable. Lettuce is sold at R$0.10 a piece, cabbage at R$0.50, cauliflower at R$0.70.

It is unquestionable that the sale price of the produce does not correspond to the reality of paying off the land debt. Legally, a loan can only be released by a financial institution if it does not compromise more than 30 per cent of the applicant's income. However, only two groups, Rei do Alface and Fazenda Akolá, actually manage to pay off the land debt. But this is not because the loan does not compromise more than 30% of their income, it is only possible because the people in these groups entered the Programme with a more privileged material structure than the others studied.

Oliveira (1982) points out that the income generated by peasant labour tends to be appropriated by financial capital through bank loans, by industrial capital through the purchase of inputs and work tools, and by commercial capital through the low prices paid for their products.

The subjection of their income to capital is not the only problem faced by the peasants studied. In all the groups, the problems began when they entered the plot, because almost none of them had basic infrastructure such as houses, electricity and running water. The amount financed for the construction of the houses was R$5,500 per beneficiary, a low amount for the purchase of materials and payment of labour. Thus, the construction of the houses was only possible with the help of the residents, as Mr A. explains.

> When we arrived here there was nothing. Everyone was worried about building a house so they could move into the plot straight away. Since money was tight, we bought almost everything we needed in the same place to get a discount. And then the people from inside and acquaintances got together (sic) to build the houses. That was the only way to build the houses for us to move in.

The practices of mutual aid and working together didn't just occur at the beginning of the group, they are recurrent, such as at harvest and planting times.

Another major problem lies in the fact that some people enter the plots without any material structure (a sum of money, transport, tools needed for production, knowledge of how to grow certain crops). This makes it even more difficult to produce for commercialisation and self-consumption. There is a clear difference between those who entered the programme with a better material condition, because they don't need to seek help from PRONAF, which makes it easier to pay off the debt.

A constant complaint in the groups is the lack of technical assistance. The technical assistance that the producers want is individual and constant, but EMATER can't provide it because there are only a few technicians, three of whom are based in

Tamarana to assist approximately 380 families. There are only three cars, with limited mileage and fuel (60 litres) per month. Due to the great distances, the technicians are unable to make at least one visit a month to each plot.

The strategy adopted by the technicians is to carry out participatory reports with everyone in each Group to raise the most urgent issues. But for this to work, the beneficiaries need to participate effectively. Another issue that hinders the work of the technicians is the high level of default by the groups, which makes it impossible to individualise and regularise the plots that have been replaced. Of the groups analysed, of those under joint guarantee, this rate reaches almost 95%. Non-payment of the land debt stems from the fact that the families are unable to pay the PRONAF, the production costs, and have a surplus to make this payment possible. The problem of default is perhaps the most striking in the groups studied, which systematically mobilises the beneficiaries, who individually or collectively seek a way out, at least temporarily. Photo 18 shows a meeting of the farmers, mediated by one of the EMATER - Tamarana technicians.

Photo 18 - Meeting of the residents of Grupo Renascer I
Source: Marcelo Campos, EMATER - Tamarana (2010)

There are always one or two plots in the Groups, especially those located in Tamarana, that have worked out very well, that manage to produce enough to pay off the land debt and provide for the survival and recreation of the families. We believe that this is due to the fact that the residents of these plots have already entered the Programme with minimum structural conditions, such as enough money to not need to resort to other agricultural credits, apart from the one already acquired with the purchase of the plot, a car, production knowledge and the tools needed to make the plot produce. However, only a minority of the farmers studied entered the plot under these conditions.

The situation of the Renascer I producers is not dissimilar to that of the other producers studied, as we will see in the course of this study.

The Renascer II and Renascer III Groups are located in the same area with a single headquarters. The reason for the Groups being separated into two separate producers' associations (Renascer II and Renascer III) is purely bureaucratic, as one of Banco da Terra's rules was a limit of 30 families per Group; there were 35 enrolled in the Programme, so they were divided into two Groups, one of 18 families, Renascer II, and the remaining 17 families formed Renascer III. Figure 9 below illustrates this.

Figura 9   - Plan of the Renascer II and Renascer III plots
Source: EMATER (2010)

Each association has its own president, vice-president, treasurer and secretary. However, the procedures of the two associations are carried out by the president of the Renascer II Association, who is on the staff of the Tamarana Rural Union, and who drew up the documents for setting up the associations.

In Tamarana, the enrolment process was different from that stipulated by the MDA, in that those interested had to set up a group and elect a representative, who would be in charge of enrolling those interested and also checking the price of the area chosen for the future installation of the plots. There, those interested in taking part in the programme had to initially join the Tamarana Rural Workers' Union. After joining, the

Union would check whether the interested party complied with the Programme's rules - having been a rural worker for at least 5 of the last 15 years, not having personal assets in excess of R$30,000.00, not owning land, or owning land considered insufficient for family reproduction. If these requirements were met, the union would issue a Declaration of Qualification. After that, the procedures were the same as those proposed by the MDA, explained in Chapter 2.

Renascer II is made up of 18 families in a total area of 101.9 hectares, with each plot covering 5.66 hectares. The purchase price of the property was R$459,999.99, and the total amount financed was R$568,779.99. The amount financed is made up of the value of the construction of 18 48m homes[2] , R$90,000, a water supply system, R$5,500, the implementation of an internal electrification network, R$7,000, the construction of internal access roads, R$3,630, notary costs, R$10,500 and topographical expenses, R$2,700.

In Renascer III, 17 families were allocated a total area of 96.8 hectares, with each plot covering 5.69 hectares. The purchase price of the property was R$439,999.56, and the total amount financed to set up the project was R$537,189.56.

Most of the residents of the Groups are from the municipalities of Londrina and Tamarana. Those from other municipalities in Paraná (Apucarana, Ortigueira) and other Brazilian states (Minas Gerais, Bahia, Pernanbuco) have lived in the region for more than 20 years.

The age range of these producers varies between 20 and 66 years, with a predominance of farmers aged between 30 and 40. The level of schooling can be considered low, as most of the producers have only completed primary school (grades 1ª to 4ª). Before joining the programme, 70% of the farmers in these groups lived in the countryside, either as employees or in partnerships. The 30 per cent who lived in the city worked underemployed as maids, porters and did occasional jobs, the so-called "odd jobs", to make ends meet.

The agricultural production of these Groups includes the cultivation of green beans, cucumbers, beans, manioc, ginger, aubergine, sweet potatoes, green corn, coffee, eucalyptus, cabbage, chard and courgettes.

It should be noted that the eucalyptus area is an initiative of the producers themselves to sell wood in the Tamarana region, and there is no contract with any timber company, not even a large paper manufacturer. The pasture areas - 38.7 hectares - are destined for this purpose due to the presence of steep slopes and rocky outcrops, which make any kind of production unfeasible. There is a 23-hectare area of woodland that is not part of the collective legal reserve. It has a very steep slope and is unsuitable for the crops grown by the groups.

Much of the produce is marketed at CEASA through middlemen, which greatly reduces the farmers' income. Some plots have a means of transporting their produce. So sometimes some plots manage to combine their produce to sell directly at CEASA, thus getting a slightly higher price for their products, as shown in Chart 8.

Table 8 - Commercialisation price of Renascer II and III Groups' production

Products	Price (Direct from CEASA)		Price (Crossing)	
Chard	R$	13,00	R$	8,00
Courgette	R$	9,00	R$	5,00
Sweet potatoes	R$	12,00	R$	5,00
Cabbage	R$	13,00	R$	7,00
Green corn	R$	7,00	R$	4,00
Pods	R$	14,00	R$	6,00
Cucumber	R$	10,00	R$	4,00
Cassava	R$	6,00	R$	3,50
Ginger	R$	15,00	R$	6,00

Source: Field surveys. June 2010.

We can see that the commercialisation prices for the middleman are almost half the price paid by CEASA. This is often one of the main factors that make it impossible to pay off the land debt.

It should be noted that the middlemen, in many cases, are the Group's own producers who, because they are in a better financial position, have a better income.

have a means of transporting their produce and are thus able to earn a higher income, while charging a percentage for transporting other farmers' products.

We can't help but consider situations in which the middleman is also a beneficiary of the Group, but doesn't have a more privileged monetary condition, but rather is unable to make his plot produce, whether due to old age, lack of resources, illness in the family, but because he owns a lorry or pick-up truck, he transports the produce of other producers to provide for his livelihood, so in this case, in our view, the middleman no longer has the image of someone who appropriates other people's income, even if this happens in practice. This happens with the beneficiaries of Grupo Brasil, which we will analyse later.

The difficulties faced by these groups, as in almost all of those studied, began when they joined the programme, as in Renascer I. They faced problems with the construction of the houses, which in many plots had to be built only with the amount financed. As the amount was low, many houses remained unfinished, some were built in wood to reduce costs, as is the case with the house shown in Photo 19 below.

Photo 19 - Precarious housing

Source: Fieldwork. June 2010.

These farmers have been in these Groups for seven years now, and according to them, the situation has improved since they joined the Programme, as we can see in the words of Mr S. from the Renascer III Group,

> "Ah! fia, tell yourself the truth, things have improved, because before we lived on other people's property. Today, even with all the difficulties, paying for the land, we're living in what's ours [...]"

It's worth pointing out that the improvement pointed out by the peasants in these groups is restricted to the acquisition of the land, to living on the plot which, after payment, will belong to them, or as Mr S. says, "what belongs to us".

We found that when some producers from Grupo Renascer I, II and III moved to the plot, due to financial difficulties, they went to work at the Fazenda Escola of the Universidade Norte do Paraná (UNOPAR), both because of its proximity and because of the possibility of earning more quickly in order to get the plot producing. These producers told us that they went to look for work outside the plot because the amount of money from the loan, which was supposed to be used to build the house and fund the first production, was insufficient. This is illustrated in the speech by Mr S., from Renascer III,

> "Until February [2010] the plot was closed, because I got a job here at UNOPAR, mowing pastures. It was registered, I had a fixed income plus a percentage of the production. I went to look for a job there, because I

> couldn't afford to run the plot. The money was only enough to build a small house with two bedrooms, a living room, kitchen and bathroom. Now I'm going to start doing something, I'm working on the land and I'm going to plant coffee."

As already mentioned, all the producers interviewed feel satisfied with their plots, although they are very worried about paying off the debt on the land and PRONAF. In these groups, the value of the annual instalment is R$4,000 (land plus PRONAF), but when asked if they would be able to pay this off, they all said no, because with the amount they receive from commercialising their produce, they wouldn't have the money needed to make this payment.

The few who have already managed to consolidate their plots and pay these instalments are unable to do so, since the financing contract was signed collectively. So if one beneficiary doesn't pay, the others - who can afford it - can't either, because the Bank only receives the total amount of the instalments for all the plots.

As such, the process of individualisation is acclaimed by those who want to pay off their debt but can't because of the joint and several guarantee. However, this is not everyone's cry, because those who are unable to pay off their debt are totally against individualisation. In this way, they feel safe, as they believe that the Bank will not take away everyone's land.

These situations are indicative of the fact that the consequences of the land credit policy do not create homogeneities that allow us to affirm its complete failure from the point of view of peasant re-creation, although this is true for the majority of cases, as has been shown so far.

3.1.5 Brazil Group

Grupo Brasil is made up of 50 families and is located on a total area of 681.78 hectares, with each plot having an area of 13.64 hectares. The purchase price of the property was R$1,184,270.00 and the total amount financed was R$1,500,000.00. Agricultural production is also geared towards olericulture. However, due to the presence of a large number of rocky outcrops in the soil and the steep slope of the land, dairy production has been favoured. This Group is divided by EMATER technicians and by the residents themselves into Brasil de Cima and Brasil de Baixo, because the area of the Group is cut by a hill, a fact evidenced in Figure 10 and Photo 20, captured from the upper part called Brasil de Cima.

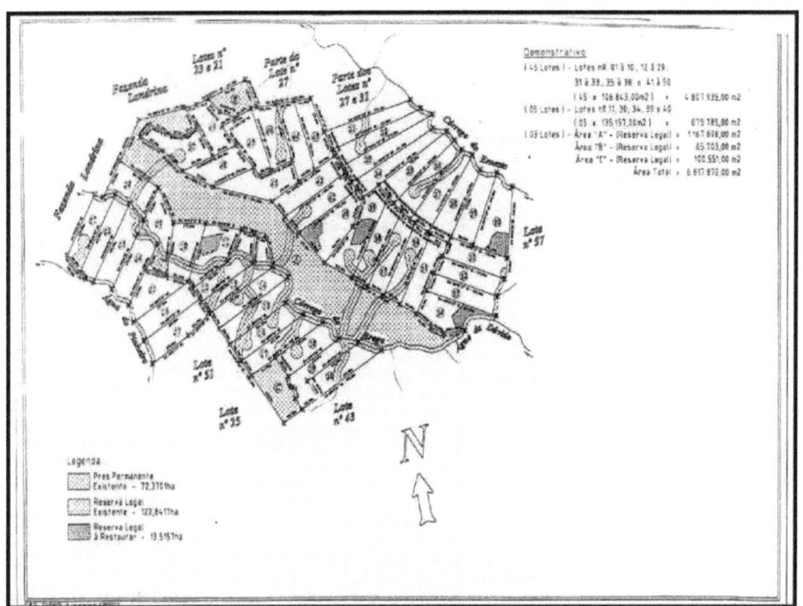

Figure 10 - Plot plan for Grupo Brasil
Source: EMATER (2010)

Photo 20 - View of the hill that cuts through Grupo Brasil, in Brasil de Cima and Brasil de Baixo.
Source: Fieldwork. July 2010.

This group, along with the Esperança Group, which we will analyse later, and the Rei do Alface Group, were not set up under the aegis of the solidarity guarantee, unlike the other groups studied. However, the Brasil Group is unique in that the owner of Fazenda Brasil, which gave rise to it, was the one who offered his employees the chance to buy the property by joining the Programme.

Mr S., who was administrator at the time, told us that

of the Treasury that,

> "he [the owner of Fazenda Brasil] came to propose it to me [joining Banco da Terra], I already wanted a piece of land to raise some dairy cows, which is what I already had here. I knew some people who also wanted it, and the union also helped organise the people. All we wanted was for the land not to be in everyone's name, as if everyone had bought just one thing. Then, as there had to be a guarantor, instead of a joint and several guarantee, there's a cross guarantee, which works like this: I'm my neighbour's guarantor and he's someone else's, but they can only be people from here, but they're not like each other. If one doesn't pay, no one can pay. Here, if the neighbour I'm guarantor for doesn't pay, the Bank [of Brazil] contacts me to find out if I can pay his debt. If neither he nor I can pay, the two of us are "seprocado[31] ". So it's almost the same as the other one, but here those who manage to pay don't get tied down by those who don't pay, because in general people chose people they knew to act as guarantors."

The fact that the land financed by the programme was paid for in cash and negotiated at market price meant that many landowners in the region sought out interested people and sold their farms via Banco da Terra. This was because, in general, the soil was already exhausted, showing low fertility, or the land was located close to settlement and encampment areas, as in the Esperança Group, causing the property to depreciate in value. It should be noted that, for the large producers in Tamarana, the presence of an MST camp or settlement close to their properties poses a risk, since in their minds, their land will be occupied and expropriated by INCRA. Therefore, selling the land before it is expropriated for social interest is much more advantageous.

As in the other groups studied, most of the members are from the municipalities of Londrina and Tamarana. Even those who are from Bahia, according to the origin identified during fieldwork, have lived in the region for at least 20 years.

We noticed during our visits that the majority of the residents in this group are aged between 51 and 61, and 21 and 31. We can consider this to be a population made up mostly of young people and adults. In families made up of older patriarchs, aged between 55 and over 61, as a rule their children don't live on the plot, either because they live in the cities of Londrina and Tamarana, or because they have a plot in the same Group or in settlements in the region.

The level of schooling can be considered low, since more than 50 per cent of the population has not completed primary school. The adult population that started primary school did not complete it. All the school-age children in the group attend state

---

[31] The word "seprocado" comes from the acronym of the Credit Protection Service (SEPROC). Therefore, those who are "seprocado" have their Individual Taxpayer Registry (CPF) blocked from any credit benefits.

or municipal schools in the municipality of Tamarana, while illiteracy is more related to the over-70s.

As mentioned earlier, most of the residents of this Group were employees of Fazenda Brasil. They were therefore salaried workers who joined the Group. According to Mr T., they were "looking for a better life, a plot of land of our own", but what was once a long-awaited dream became a latent nuisance, because the R$1,500,000.00 was only for the purchase of the land, while only R$12,000.00 was allocated to each family for the technical project, via PRONAF. This was also the case for all the groups studied, with the exception of the King of Lettuce Group. Only R$3,800.00 (included in the total value of the financing) was earmarked for the construction of the houses.

The technical or economic viability project carried out by EMATER - Tamarana states that the beneficiaries should produce vegetables (especially potatoes, yams, lettuce and cabbage), coffee, dairy and beef cattle. We were able to observe in the field that this project has been fulfilled, with the exception of the production of vegetables and plots where cereals are produced (corn, wheat and soya). Only one plot we visited, shown in Photo 21 below, does not produce anything for sale.

Photo 21 - Housing conditions of elderly people covered by the Programme.
Source: Fieldwork. July (2010)

This is because the residents of this plot, a 67-year-old couple, are unable to produce, because according to them:

"We're already old, we can no longer manage to produce to sell. Our children all live in the city, some in Londrina, others in Tamarana. We produce to eat, we plant manioc, chayote, some peppers and coffee. We manage to survive because T. takes the neighbours' produce to CEASA. When we manage to produce a little more, he sells it at CEASA. But that's very difficult to do. We'd like to sell more, but then we'll have to pay someone to help. Vixi, if we pay someone, we won't have any money to survive. So the deal is to carry on as you are, because without paying someone to help, we can't pay off the debt properly [...]" (sic)

In all the other plots visited, there is production for

Most of the plots are dedicated to dairy and beef farming. Most of the milk producers deliver their production for R$0.60 a litre to CATIVA, so they are part of the cooperation system, like the producers in Renascer II and III.

In July 2010, beef cattle producers were selling at R$65.00 an arroba, with a regular rate of 25 head per month. It should be noted that the beef cattle producing plots visited are two in number and belong to the same family. As a result, they are able to have a larger area of pasture and greater production. This family told us that the cattle are sold to meatpackers in the region.

In this group, as well as the other groups studied in Tamarana, there were some families who managed to consolidate their production and have production geared towards commercialisation, without the need for middlemen or even having to join the integration system. One of these plots is considered by EMATER technicians to be a reference property, which is the plot of Mr S., the former administrator of Fazenda Brasil, as shown in Photo 22 below.

Photo 22 - Reference property.

Source: Marcelo Campos, EMATER - Tamarana (2010)

The producer of the reference plot told us that he managed to reach this level of production because he had enough money to invest in the plot. He considers himself lucky to have been able to buy everything he needed to process the milk.

According to Mr S., "we've been stuck here for 10 years, we [he and his wife] were lucky. The cattle I have today have nothing to do with the money from the project. It was a lot of intelligence and willingness to work."

It should be noted that the house on this plot was the former headquarters of Fazenda Brasil, so it was not built with the money financed for this purpose.

The herd on this plot is made up of mixed-breed cattle, as shown in Photo 23 below. The choice of mixed-breed cattle is due to the fact that the costs of top genetics, even though they produce more milk per day, are higher, as are the risks of disease. Another factor that leads to this choice is the lower feed costs for mixed-breed cattle than for pedigree cattle.

Photo 23 - Mestizo cattle
Source: Marcelo Campos, EMATER - Tamarana (2010)

Sometimes the option of mixed-breed cattle is not well regarded by the extension technicians, because the rationale imposed on them (by the Programme) is that of the market, i.e. the technicians must guide the beneficiary to produce more in order to get enough income to pay off the debt. However, this logic often doesn't correspond to reality or even to the rationality of these farmers.

We can consider that peasant rationality is one of security. On several occasions, peasants choose to produce less, as in the case of this plot of land, which favoured the breeding of mestizo cattle over pedigree cattle, in order to maintain the security of having regular production, to guarantee their livelihood and the payment of the land debt and any agricultural loans.

In this lot, the milk is not sold to any co-operative or dairy industry. The milk is processed on the plot itself and sold to final consumers in the municipalities of Londrina and Tamarana. In this way, the producers get a better price, since they can dictate the selling price. The dairy products produced on the plot are basically cheese, dulce de leche and curd cheese, which are regularly sold to the end consumer. The photo below (Photo 24) shows the stock of produce to be commercialised.

Photo 24 - Milk derivatives.
Source: Marcelo Campos, EMATER - Tamarana (2010)

Alongside this production, there are areas of orchard, manioc, vegetable garden and pig and chicken farms on the plot for self-consumption.

It's worth noting that sometimes, when milk production is lower, Mr S. buys milk from some neighbours so that he can benefit from it.

The irrigation system on this plot is carried out by collecting water from a body of water formed by a waterfall on the plot, which can be seen in Photo 25 below.

In short, these are the expressions of the limits of possibilities for peasant recreation identified in Grupo Brasil.

Photo 25 - Waterfall.
Source: Marcelo Campos, EMATER - Tamarana (2010)

3.1.6 Hope Group

The Esperança group, made up of 8 families, is located on a total area of 82 hectares, 10.25 hectares per plot. The amount financed was R$284,640.00. As a rule, it is the soil and morphological conditions that have led to larger plots than in the other projects, as the terrain has a steep slope and there are also intense rocky outcrops.

We should also point out that this, along with the Brazil group, are one of the few Banco da Terra projects that were not set up under the aegis of the solidarity guarantee and that have individual plots, i.e. it was not necessary to organise an association to form the group.

In the Esperança group, we found only a few producers who effectively make a living from agricultural production, with a predominance of vegetables, oranges and coffee.

In this group, we came across some people who keep the plot as a leisure farm for the weekends, such as a plot belonging to a pharmacist who lives in the municipality of Londrina.

The process of buying this property was similar to that of Fazenda Brasil. The owner of the old farm offered it to Banco do Brasil so that he could sell it via Banco da Terra. This was because, according to the owner, there were two aggravating factors to selling the farm without the mediation of the Programme. Firstly, the location of the Serraria settlement, which hosts frequent meetings of MST activists, next to the property, and the MST camp, located at the entrance to the property. As already mentioned, the location of settlements and encampments bothers the landowners in the neighbourhood.

As with the other groups studied in Tamarana, Esperança's producers come from rural areas in the municipalities of Londrina and Tamarana. Those from other municipalities in Paraná have lived in the region for more than 10 years. The majority of the population in this group is middle-aged (40 to 60 years old). The level of education varies with the age of the residents. There are plots where the level of education ranges from illiteracy to incomplete secondary education, as well as one plot where almost all the residents have completed secondary education and one of the children has completed higher education and postgraduate studies.

In this group, the families are not large and generally consist of five people. There is one plot where only women live: the widowed grandmother, two divorced daughters and two granddaughters. This is an unusual situation among the plots studied, which does not directly interfere with the work on the plot.

We were very impressed when we arrived at the plot and were met by a very simple 76-year-old woman with the marks of farm work on her face and hands. We learnt that only she, her daughters and granddaughters, who had always lived in the countryside, lived on this plot. Before joining the Programme, Mrs R., the matriarch, lived with one of her daughters on a farm as sharecroppers and the other daughter lived with her husband, also as sharecroppers in another area in Tamarana. After the divorce, she went to live with her mother and sister.

There are few crops on this plot, as the soil is of low fertility, with a steep slope and rocky outcrops, not unlike the conditions of the other Groups already presented. The plot produces maize, beans and broom. However, little of what is produced is destined for the market, as they have to pay for transport to CEASA, and on several occasions, the price obtained in the market does not cover the price of production. They don't have the money to diversify their production with crops that grow better in this type of terrain and have greater added value, such as citrus fruits, passion fruit, pasture, among others. To get a better income, they rent part of their plot to Mr L., a beneficiary of the same group.

Mr L.'s plot is home to five people: him, his wife, two sons and a daughter. Mr L finished high school and is an agricultural technician. His wife didn't finish high school, his male children have finished high school and are applying for higher or technical education, and his daughter has a degree in pedagogy and is doing postgraduate studies.

The labour force on the plot is essentially family. With the exception of the daughter, who studies and works in Londrina, but occasionally helps out with chores on the plot, all the residents work effectively in production.

Mr L. has a very interesting story of life and work on the land. Born in Palotina, PR, he has always worked in the fields, on his father's farm, growing soya beans and sunflower seeds, which were destined for the vegetable oil industry. After several crop failures, his indebted father lost the farm to the bank. So, in the 1970s, he moved with his wife to Sinop - MT, with hopes pinned on the promise of cheap and fertile land, like many people from Pará who migrated to the north of the country at the height of the coffee crisis.

After living in Mato Grosso for twenty years, he and his family migrated to Bolivia, where they lived for four years, but were expelled in one of the waves of expulsions of foreigners who owned land in the country. So he went to work as a tractor driver on a farm in Tamarana, when he learnt of a family who wanted to leave their plot in Grupo Esperança, and so he decided to take their place.

Mr L told us that:

> "Life has always been very difficult, first we lost our farm to the bank, then I went to Mato Grosso and we stayed there for a while, my children were all born there. Then it was right to go to Bolivia, we stayed there for four years, but then we were expelled without the right to the land. After that we came here to Tamarana, I went to work on a farm as a tractor driver. Now everything is better, I have my land. [...] I don't think it's difficult to pay for the land, it's expensive, there's no denying that, but with effort and work we manage to pay for it. [...] I don't think I have much trouble paying for it, because I didn't get here like a lot of people, with one hand in front of me and the other behind[32] . I had a bit of money, a cart, I was already an agricultural technician, so technical assistance was never lacking, and I also worked hard to make everything work here."

This statement proves our thesis that peasants only have the chance to recreate themselves within Banco da Terra when they enter the programme with minimal structural conditions.

This plot produces a variety of crops, fruit such as passion fruit and oranges, chilli, yams, corn, coffee (self-consumption), courgettes, soya beans, in an area leased from Mrs R.'s plot, and also pasture for raising dairy cattle. This pasture area is

---

[32] The expression with one hand in front and the other behind means that the people arrived at the lots with no money.

also leased from the plot belonging to the pharmacist from Londrina, and there is also an area of eucalyptus that is sold for firewood.

As well as working on the plot, Mr L also earns an income from his work as a tractor driver for neighbouring plots. He told us that when he's working outside the plot, it's his children and wife who take care of the land.

The produce from this plot is sold at CEASA and transported by Mr L, who also transports the produce of his neighbours. The milk is delivered to Cativa.

We can see that this plot is consolidated, with a few already mentioned. However, this plot is the exception to the rule, as it was one of the few peasants who told us that there was little difficulty in paying off the land debt.

As you can see, the Esperança Group escapes the rule of the land banks studied and, to an equal extent, presents the limits that are inherent to the political conception of agrarian reform in the hands of the market, which does not invalidate the few exceptions in which the process of differentiated peasant recreation can be glimpsed.

3.2 The Land Bank in Londrina and Tamarana: Limits and Possibilities

During the course of this research, we spent six months in the field, trying to understand a little of the reality of the 110 families studied and, in all the groups, we found similar realities. There are structural features of the Programme that are similar to all of them, the most striking being the small size of the plots. In the municipalities of Londrina and Tamarana, the fiscal module according to Law 8.629 of 25 February 1993 is 12 hectares, but we found that none of the groups had plot sizes equal to or greater than the municipal fiscal module.

Another limiting factor of Banco da Terra is that none of the financing contracted by the producers is allocated to the production project, but only to the infrastructural conditions. This means that producers have to resort to other land credit lines such as PRONAF. As a result, they end up in even more debt, since the Land Bank Programme's interest rate for the region is 12% per year and the instalments average R$4,000.00.

Producers' indebtedness often makes it impossible for them to take out other loans to produce, since peasants whose names are on SERASA cannot access any kind of financing, which can make it impossible for them to recreate their livelihoods, and they may abandon or sell their plots. This indebtedness leads to a high level of default among the beneficiaries of the Groups studied, which reaches 90 per cent. Beneficiaries who manage to pay off the debt on the land are rare and express a different situation in which they already had the minimum conditions to recreate themselves before joining the programme.

Another similarity identified between the groups was the way they were set up. With the exception of the King of Lettuce, all the groups were set up with the intervention of the Tamarana Rural Workers' Union. As already mentioned, almost all of the families studied lived in municipalities and districts in the Londrina and Tamarana region and were of rural origin (tenants or salaried rural workers).

In an interview with the President of the Tamarana Rural Workers' Union, in order to investigate the mediators' role in consolidating the groups, she told us that when the Programme came into force in the region, everyone (in the Union) was very hopeful, given the promise of making it possible to buy land and consolidate family farming. Because of this and the intense advertising and facilitation by the state, many people came to the union interested in taking part.

The president told us that the Groups were formed in the following way: interested people went to the Union to sign up, some of whom already had the Group formed, but many of the people who signed up were unable to take part in the Programme because they had irregular documents. After registration was finalised, the Groups were formed almost at random and then the process of choosing the area and negotiating the purchase began.

When asked if any landowners in the region had offered properties to be sold via Banco da Terra, the answer was negative. However, some of the interviewees said they had been approached by landowners interested in negotiating their land through the programme, such as Grupo Brasil. This shows that local authorities have become involved in the programme and the number of Land Bank Groups, especially in the municipality of Tamarana.

It should be pointed out that, as a rule, the groups are located in areas far from the marketing channels, with low-fertility soils, rocky outcrops and steep slopes (especially in Tamarana), as shown in Figure 3. This leads to difficulties in production and is one of the variables to be taken into account when making plots unviable.

Finally, the principle of the joint guarantee process is also a limiting factor for the re-creation of some of the families studied, because families with enough money to pay into the[33] programme find it very difficult to pay the loan instalment, since

---

[33] In the joint guarantee process, each member of the group acts as guarantor for the others. In the case of Banco da Terra, when one "beneficiary" doesn't pay the annual instalment of the loan, all the others can't make the payment, because as the property is in the name of the Association or Group, the bank only receives the total amount of the loan and not what is due to each

the bank only receives the total debt of the Group and not per beneficiary. This means that if one family doesn't pay, no one else in the group will be able to pay, leaving them in default and without access to finance.

On this subject, the president of the Akolá Farm Producers' Association told us that in March 2010, he had already asked the Municipal Rural Development Council (CMDR) to carry out the process of replacing some plots that the "beneficiaries" had given up on and the process of individualising the financing debts. Despite having been voted on and approved at an extraordinary meeting of the CMDR on 11 March 2010, these processes have not yet been completed, because in order for the replacements and individualisation to be carried out, various documents need to be submitted, which will go through different institutions at municipal, state and federal level, such as EMATER, SEAB, MDA and Banco do Brasil. This slows down the process.

The production of the plots of the Groups studied is a good indicator of the contradictory possibilities experienced by peasants. There are those producers who already had minimal conditions and managed to consolidate themselves on the plot, producing for self-consumption and for commercialisation. This has enabled them to pay off the land debt, despite all the adversities that are common to all, such as production costs, transport, family support, location, soil quality, among others. However, the others tend to perish as autonomous farmers on their plots, both because of the Programme's structural problems and because of impediments within each family.

The condition in which peasants are subjugated also poses difficulties; we can see that the subjugation of these producers, in all the Groups, is not only in the area of product commercialisation. They are subjugated by financial capital, by the debt from the purchase of the plot, by industrial capital, when they buy inputs and seeds that are constantly increasing and often don't match the price they get on the market.

Faced with all these problems, the families try to find ways not to lose their dream land. Many, for lack of conditions, stop producing for sale on their plot to work elsewhere as rural wage earners. This is so that they can pay for the land; others act as middlemen, and some try to sell their produce in ways other than through CEASA, such as at fairs or on the plot itself.

The way in which the programme has been structured reveals its inability to prove itself as a viable agrarian reform model, because even though it facilitates the purchase of land, it often fails to provide the families benefiting from it

---

"beneficiary".

with the conditions to survive, thus confirming that the market is not capable of solving the problem of the Brazilian agrarian question.

CHAPTER 4

FINAL CONSIDERATIONS

The way in which the Western Christian appropriation of Brazilian territory took place, combined with the Portuguese Monarchy's lack of control over the limits of concessions and possessions, culminated in the emergence of the latifundia.

The maintenance of this latifundist structure was reaffirmed through various laws and government programmes, such as the Land Law of 1850, the Land Statute of 1964 and the land credit policy of the 1990s and 2000s. Each of these acts allowed for the renewal of the existing power pact between the state and large landowners, thus reinforcing the class monopoly over land.

The concentration of land, which was an obstacle to the development of capitalism and complete modernisation in central capitalist countries, in Brazil took on a complementary character to the structures of expanded accumulation, in the context of the metamorphosis of industrial and urban capitalists into landowners, which sealed the alliance between land and capital, in the words of Martins (1994).

These actions were combined with strategies to selectively alleviate tensions through compensatory social policies which, instead of acting as corrective public policies, culminated in the strengthening of the latifundia, which in itself was capable of suppressing mobilisations and struggles for the democratisation of access to land.

This context motivated this research, which sought to analyse the Banco da Terra Programme and its territorial developments in the municipalities of Londrina and Tamarana - PR. The proposed territorial section is justified by the fact that Paraná is one of the main states in which the Programme operates, which in turn is already a product of the mitigating actions on agrarian conflicts mentioned above, in view of the occurrence of territories of peasant resistance, such as the encampments and settlements that have a marked presence in Tamarana, a newly constituted municipality due to the dismemberment of Londrina's political-administrative domains.

The analysis of the Banco da Terra Programme has shown that it is not an isolated spatio-temporal action, but rather part of the neoliberal structural adjustment policy advocated by the World Bank and introduced in Brazil with the backing of the state.

In general terms, this policy was extended to rural areas in most underdeveloped countries which, like Brazil, had been experiencing conflicts and peasant uprisings motivated by the ban on labour land. Hence the lines of financing for land purchases by landless rural workers or those with insufficient land to survive.

As such, it can be said that its primary objective was to contain tensions in the countryside and bring land policy into line with free market principles, even though the discourse indicated the intention of promoting well-being in the countryside and the consolidation of farmers on the land, as can be seen from the following fragment, taken from the Banco da Terra Programme's Operations Manual (2000), which states that the aim is to "promote land reorganisation and rural settlement actions, with a view to strengthening family farming in conjunction with the beneficiary rural communities and in line with the state's agricultural and rural development policy".

The gap between the announced aims and the results achieved so far was evident in this work, because in all the Groups studied, most of the plots did not meet the minimum structural conditions to produce according to the precepts of the agrarian legislation in force in Brazil, which advocates access to a piece of land capable of ensuring the satisfaction of the needs of existence and economic progress for the family living on it.

Instead of enforcing its basic prerogative of promoting land justice through the distribution of vacant land, which according to Oliveira (2003) corresponds to approximately half of the country's territorial domains, the Brazilian government opted for the so-called market agrarian reform, subsidising the land business.

Despite the credits granted to farmers, in practice it was private landowners who benefited most from the programme, as they were paid cash for land that was generally overpriced, many of which had dubious legal titles.

What's more, one would not expect this problem, which is so latent in Brazilian society, to be solved just by granting credit, a fact made clear by this research, given that only a few plots have been consolidated, with production at levels compatible with the minimum demands of the families involved. It should be noted that in all these cases, when the farmers joined the programme, they already had the minimum structural conditions to produce, including some money, equipment, machinery and sometimes even vehicles. On the other hand, the research did not identify a single case of success, within the parameters described above, among those who were strictly dependent on the resources made available to establish themselves on the plot.

This, which already appears to be a structural shortcoming of the programme, was exacerbated by mistaken technical-political prescriptions, since it was idealised that producers would be fully integrated into the market, as if they could, by decree, compete for competitive positions under symmetrical conditions with the others already established.

In short, it is this concept that explains the rule that obliges them to comply with the so-called economic viability, drawn up by technicians who define which products should be grown. In all cases, the parameter identified was crops with high levels of profitability, given the conditions of scale. However, as a rule, the beneficiaries' experience in the activity, the availability of tools and machinery, the appropriate balance between intensiveness and available family labour, distance and accessibility to the consumer market, among others, were not taken into account.

Thus, even though it provides access to land, Banco da Terra, as well as all the other programmes that favour the process of financing the purchase of rural properties over land redistribution, do not guarantee beneficiary families income conditions that ensure both the payment of debts and a livelihood.

We conclude that land credit programmes along the lines of Banco da Terra cannot be assimilated as a model of agrarian reform, since they are above all strategies for collecting capitalised land revenue by looting the public purse, which penalises not only landless peasants, but society in general, which, as well as having part of the resources diverted from essential sectors, is faced with a deepening of the structure that favours the rapine of land revenue that weighs indiscriminately on it.

REFERENCES

2,4D [DIMETHYLAMINE SALT OF (2,4-DICHLOROPHENOXY) ACETIC ACID]. Nortox SA. Arapongas: Nortox AS, 2010. Available at: <http://www.nortox.com.br/imagens/produtos/24d_bula.pdf>. Accessed on: 19 Nov. 2010.

ALMEIDA, Lúcio Flávio Rodrigues de. Social struggles and the national question in Latin America: some reflections. **Lutas sociais**, n. 17/18, p. 64-67, Jul/Dec 2006 / Jan/Jun 2007.

ALMEIDA, Rosemeire Aparecida de. **(Re)criação do campesinato, identidade e distinção**: a luta pela terra e o habitus de classe. São Paulo: UNESP, 2006.

ASSOCIATION OF TOBACCO GROWERS OF BRAZIL - AFUBRA. **Table of tobacco prices for the 2009/2010 harvest**. available at: <http://www.afubra.com.br/principal.php?acao=conteudo&conteudo_id=268&i_id=1&u_id=1>. Accessed on: 19 Nov. 2010.

BINSWANGER, Hans. **Market-assisted land reform**: the World Bank's new approach. Translated by: MAIA, José Nelson Bessa. [s.n.t.] [1996?].

BONATO, Amadeu A. **Family farming and the tobacco chain in Brazil.** EMATER, Department of Rural Socio-Economic Studies DESER. Available at: <http://www.emater.pr.gov.br/arquivos/File/Seminarios/08Cadeiadofumo.pdf>. Accessed on: 19 Nov. 2010.

BRAZIL. **Law No. 601, of 18 September 1850.** Provides for the Empire's vacant lands. Rio de Janeiro: General Assembly, 1850. Available at: <http://www.planalto.gov.br/ccivil_03/Leis/L0601-1850.htm>. Accessed on: 19 Nov. 2010.

BRAZIL. **Constitution of the Federative Republic of Brazil.** Brasília: Federal Senate, 1988.

BRAZIL. **Law No. 8.692, of 25 February 1993.** Provides for the regulation of the constitutional provisions relating to agrarian reform, set out in Chapter III, Title VII, of the Federal Constitution. Brasília: General Assembly, 1993. Available at: < http://www.planalto.gov.br/ccivil_03/leis/l8629.htm>. Accessed on: 20 Feb. 2010.

BRAZIL. FEDERAL GOVERNMENT. **Banco da Terra Programme** operations manual. Brasília, DF, 2000. Available at <http://www.incra.gov.br>. Accessed on 25/02/2009.

CHAYANOV, Alexander Von. **The organisation of the peasant economic unit.** Buenos Aires: New Vision, 1974.

CASANOVA JR., Guilherme. Despite projects and incentives, family farmers feel isolated. **Jornal de Londrina**, 13 Dec. 2009. Available at:

<http://www.jornaldelondrina.com.br/edicadododia/conteudo.phtml?tl=1&id=954054&ti t =Despite-projects-and-incentives-family-farmer-says-he's-isolated>. Accessed on: 19 Nov. 2010.

FILGUEIRA, Fernando Antonio Reis. **Novo manual de olericultura**: agrotecnologia moderna na produção e comercialização de hortaliças. Viçosa: UFV, 2000.

FURTADO, Celso. **Economic formation of Brazil.** 34 ed. São Paulo: Companhia das Letras, 2007.

KAUTSKY, Karl. **The agrarian question.** 3.ed. São Paulo: Proposta, 1980.

LÊNIN, Vladimir I. **Capitalism and Agriculture in the United States of America**: New Data on the Laws of the Development of Capitalism in Agriculture. São Paulo: Brasil Debates, 1980.

LENIN. Vladimir I. **Capitalist Development in Russia.** São Paulo: Abril Cultural, 1982.

LESSA, Antonio Carlos. **15 years of economic policy.** 2.ed. São Paulo: Brasiliense, 1981.

MAIOR, Mylena Fiori C.; SALVO, Maria Paola de. **End of the land bank.** Available at: <http://www.midiaindependente.org/eo/blue/2003/02/248189.shtml>. Accessed on: 12 March 2008.

MARTINS. José de Souza. **A militarização da questão agrária**: terra e poder, o problema da terra na crise política. Petrópolis: Vozes, 1984.

MARTINS, José de Souza. **Os camponeses e a política no Brasil**: as lutas sociais no campo e seu lugar no processo político. Petrópolis - RJ: Vozes, 1986.

MARTINS, José de Souza. **The captivity of the earth**. 4.ed. São Paulo: Hucitec, 1990.

MARTINS, José de Souza. **O poder do atraso**: ensaios de sociologia da história lenta. São Paulo: Hucitec, 1994.

MARTINS, José de Souza. **O poder do atraso**: ensaios de sociologia da história lenta. 2.ed. São Paulo: Hucitec, 1999.

MARTINS, Mônica Dias. Learning to participate. In: ___________ . **The World Bank and land**: offensive and resistance in Latin America, Africa and Asia. São Paulo: Viramundo, 2004. p.61-73.

MARX, Karl. **Capital**. Rio de Janeiro: Civilização Brasileira, 1974 (Great Masters of Thought; v. 6)

MEDEIROS, Leonilde Servolo de. **Reforma agrária no Brasil**: história e actualidade da luta pela terra. São Paulo: Perseu Abramo, 2003.

MENDONÇA, Maria Luisa; RESENDE, Marcelo. Agrarian counter-reform in Brazil. In: MARTINS, Mônica Dias. **The World Bank and land**: offensive and resistance in Latin America, Africa and Asia. São Paulo: Viramundo, 2004. p.75-79.

MRTVI, Paulo. Community walk. **O homem e a terra,** Curitiba, n. 36, Dec. 2009. Available at: <http://www.emater.pr.gov.br/arquivos/File/Comunicacao/Jornais/HTEletronico/HTEn 36.pdf>. Accessed on: 19 Nov. 2010.

NABARRO, Sérgio A. **The Land Bank in Tamarana - PR**: the case of the Renascer II and III Groups. 2007. Monograph (BA in Geography) - State University of Londrina, Londrina.

OLIVEIRA, Alexandra Maria de. The World Bank's agrarian counter-reform policy in Ceará. **GEOUSP: espaço e tempo,** São Paulo, n. 19, p. 151-175, 2006.

OLIVEIRA, Ariovaldo Umbelino de. The **agrarian question and the right to land**: a movement in defence of social rights - Brasília. Available at: <http://www.direitosociais.org.br/_arquivos/2010/343_questaoagrariaparte1.pdf>. Accessed on: 19 Nov. 2010.

OLIVEIRA, Ariovaldo Umbelino de. **"Me engana que eu gosto"**: the failure to update land productivity indices under the Lula government. Available at: <http://www.direitos.org.br/index.php?option=com_content&task=view&id=2968&Ite m id=2>. Accessed on: 19 Nov. 2010.

OLIVEIRA, Ariovaldo Umbelino de. **The geography of struggles in the countryside**. 3.ed. São Paulo: Contexto, 1990.

OLIVEIRA, Ariovaldo Umbelino de. Barbárie e modernidade: as transformações no campo e o agronegócio no Brasil. **Terra Livre,** ano 19, v. 2, n. 21, p. 113-156, jul/dez.

2003.

OLIVEIRA, Ariovaldo Umbelino de. **Capitalist mode of production and agriculture**. 4.ed. São Paulo: Ática, 1995.

OLIVEIRA, Ariovaldo Umbelino de. **Capitalist mode of production, agriculture and agrarian reform**. São Paulo: Labur Edições, 2007.

OLIVEIRA, Ariovaldo Umbelino de. Reflexões sobre o imperialismo: a incorporação do Brasil ao capitalismo internacional. **Boletim Paulista de Geografia**, São Paulo, n. 59, p. 59-114, 1982.

OLIVEIRA, Ariovaldo Umbelino. Agrarian Geography and recent territorial transformations in the Brazilian countryside. In: Carlos, Ana Fani Alessandri (Org.) **Novos caminhos da Geografia**. São Paulo: Contexto, 2002.
OLIVEIRA, Ariovaldo Umbelino de. **Peasant Agriculture in Brazil**. São Paulo: Contexto, 1991.

PAULINO, Eliane Tomiasi. **Towards a geography of peasants**. São Paulo: UNESP, 2006.

PAULINO, Eliane Tomiasi FABRINI, João Edmilson; (Org.). **Campesinato e territórios em disputa**. São Paulo: Expressão Popular, 2008.

PAULINO, Eliane Tomiasi. The agrarian question and geography teaching: a necessary debate. In: KATUTA, Ângela Massumi et al. **Geografia e mídia impressa**. Londrina: Moriá, 2009a. p. 61-86.

PAULINO, Eliane Tomiasi. **Production of own seeds:** a fruitful encounter between science and peasant knowledge in the Londrina region - Brazil. 2009b Available at: <http://egal2009.easyplanners.info/programaExtendido.php?sala_=A%20-%2002&dia_=DOMINGO_AREAS_6_7_8#>. Accessed on: 29 Mar 2010.

PAULINO, Eliane Tomiasi; ALMEIDA Rosimeire Aparecida de. **Terra e território**: a questão camponesa no capitalismo. São Paulo: Expressão Popular, 2010.

PEREIRA, João Márcio Mendes. The political-ideological dispute between redistributive land reform and the World Bank's market-based land reform model (19942005). **Sociedade e Estado,** Brasília, v. 20, n. 3, p. 611-646 Sep/Dec 2005.

PEREIRA, João Márcio Mendes. The World Bank's agrarian policy in question. **Estudos Avançados**, São Paulo, v. 20, n. 57, Aug. 2006. Available at: <http://www.scielo.br/scielo.php?script=sci_arttext&pid=S0103-40142006000200024&lng=en&nrm=iso>. Accessed on: 19 Nov. 2010.

PEREIRA, João Márcio Mendes; SAUER, Sérgio (Org). **Capturing the land**: World Bank, neoliberal land policies and market agrarian reform. São Paulo: Expressão Popular, 2006.

PLOEG, J. D. van der. **Peasants and food empires:** struggles for autonomy and sustainability in the age of globalisation. Porto Alegre: UFRGS, 2008.

RAMOS FILHO, Eraldo da Silva. The **Current Agrarian Question**: Sergipe as a reference for a comparative study of agrarian reform and market agrarian reform policies (2003 - 2006). 2008. Thesis (PhD in Geography) - Universidade Estadual Paulista, Presidente Prudente.

RODRIGUES, Lúcio Flávio de Almeida. Social struggles and national issues in Latin America: some reflections. **Lutas Sociais.** v. 17/18 p. 64-77, jan/jul, 2007.

ROSSET, Peter. The good, the bad and the ugly: World Bank land policy. In: MARTINS, Mônica Dias. **The World Bank and land**: offensive and resistance in Latin America, Africa and Asia. São Paulo: Viramundo, 2004. p.16-24.

SALLUM JÚNIOR, Brasílio. **Capitalism and Coffee Growing**. West São Paulo: 18881930. São Paulo: Duas Cidades, 1982

SAUER, Sérgio. Land as a banknote: a study on "market-based agrarian reform". In: MARTINS, Mônica Dias. **The World Bank and land**: offensive and resistance in Latin America, Africa and Asia. São Paulo: Viramundo, 2004. p.16-24.

SCIARRA, Eduardo. **Planalto gives in to ruralists on productivity index**. Available at: <http://eduardosciarra.com.br/planalto-cede-a-ruralistas-em-indice- de-productividade/>. Accessed on: 19 Nov. 2010.

SECRETARY OF STATE FOR AGRICULTURE AND SUPPLY - SEAB.
Hit the right target: eliminate drift in pesticide spraying. Information leaflet. Available at :
<http://www.seab.pr.gov.br/arquivos/File/defis/educacao/alvo1.pdf>. Accessed on: 19 Nov. 2010.

SHANIN, Teodor. The definition of peasant: conceptualisations and de-conceptualisations - the old and the new in a Marxist discussion. **Revista Nera**, Presidente Prudente, Ano 8, n. 7, p. 1- 21, Jun/Dec 2005.

SHANIN, Teodor. **Peasant lessons**. In: ____________ PAULINO, Eliane Tomiasi FABRINI, João Edmilson; (Org.) Campesinato e territórios em disputa. São Paulo: Expressão Popular, 2008. p. 23-47.

SILVA, Lígia Osorio. **Vacant lands and latifundia**: effects of the Land Law 1850. Campinas: Editora da UNICAMP, 1996

SILVA, Lígia Osorio. **Vacant lands and latifundia**: effects of the Land Law 1850. 2ed. Campinas: Editora da UNICAMP, 2008

SORJ, Bernardo. **State and classes in Brazilian agriculture**. Rio de Janeiro: Editora Guanabara, 1986.

STÉDILE, João Pedro (Org.) **A questão agrária no Brasil**: programas de reforma agrária 1946-2003. São Paulo: Expressão Popular, 2005.

TSUKAMOTO, Ruth Youko; ASSARI, Alice Yatiyo. **Rural settlements and economic and spatial reorganisation - Tamarana - PR**. Available at: <http://www4.fct.unesp.br/nera/publicacoes/singa2005/Trabalhos/Artigos/Ruth%20Yo uko%20Tsukamoto.pdf> . Accessed on 21/03/2009.

CHAPTER 5

APPENDIX

APPENDIX A

Questionnaire

1. Life story.

1.1.   Age

1.2.   Marital status

1.3.   What level of education do you have?

1.4.   Do you have children? How many? Age?

1.5.   Do your children study? What grade? Which school? Is the school in the district or in the city of Londrina? What means of transport do they use?

1.6.   Do you have a family member who works or receives a pension?

1.8.   How much, in proportional terms, does this external income represent?

1.9.   Is it a fixed income, or does it vary throughout the year? (If it varies, take the annual average)

1.9.1.   If this monetary inflow were to cease, would survival on earth be jeopardised?

1.10.   Have you always lived in the countryside? (Did you own the land? Size of area farmed)

1.11.   What activities were carried out? Which crops? Was there commercial farming of poultry, pigs, cattle, etc.?

1.12.   What did you do before joining the Group?

1.13.   If you don't have a tradition in agriculture: What prompted you to join the Group?

1.14.   How long have you been part of the Group?

1.15.   Are you part of the original group? If not, what is the order of occupation?

1.16.   Have you ever been a camper or settler? If so, why did you leave the camp/settlement and join the Group?

1.17.   How did you hear about Banco da Terra?

2.   The Group.

2.1.   How was the Group formed?

2.2.   How were the people chosen to be part of the Group? What criteria were used?

2.3.   Who owned the land?

2.4.   ow much was financed? Have you just paid off the mortgage? How much time is left?

2.4.1.   How much is the annual instalment? How much has already been paid? How difficult is it to pay? Is the group under a joint and several guarantee? Have there been any problems? What is the percentage of defaulters in the group?

2.4.2.   What are your future projects?

2.5.   Is the land already titled?

2.6.   Since you came to the Group, has your life improved, worsened or stayed the same?

2.7.   Is there an internal activity that socially integrates the project's residents (rosaries, festivals, etc.)? There are institutional measures present in the "settlement".

2.8.   How is the process of individualising plots going? Are you in favour or against?

2.9.   Are you afraid of losing your plot if you don't pay your mortgage instalments?

2.10.   What do you think of the Land Bank Programme? Is it better than where you were before?

3.   Plots.

3.1.   Are all the plots the same size?

3.2.   How big is your plot?

3.3.   What is produced on your plot? Is there crop rotation? Are there any crops that have a short cycle?

3.4.   What is the area occupied by each crop produced on your plot?

3.5.   What is the soil like?

3.6.   What is produced for self-consumption? How much of the plot do you occupy?

3.7.   How many people work on the plot? (How many are family members?) working hours.

3.7.1.   Do you have a paid helper? Permanent or seasonal?

3.8.   Do all the plots have water? Is it well water? Is it communal?

3.9.   As for the plot's infrastructure:

3.9.1.   Production storage shed

3.9.2.   water mine

3.9.3.   lorry/utility

3.9.4.   electric light

4.   Production.

4.1.   What equipment is used in production? Do you have all of them? If not, how do you manage production?

4.2.   Is there a regularity in production?(month/year)

4.3.   How much do you sell your produce for?

4.4.   Have you suffered any crop failures since you joined the Group (frost, drought, rain, etc.)?

4.5.   Are the members' productions put together for the market?

4.6.   What do you do with what you can't sell? When you lose?

4.7.   Which crops do you use pesticides on? How often do you apply them? Where do you buy them? How do you pay for them?

4.8.   Has anyone in your family ever had problems using poison?

4.9.   What do you do with the packaging?

4.10.   Do you use any natural pest control practices and/or organic fertilisation on your crops?

4.11.   Do you have technical assistance? How often? How often? Is this a high cost for you?

4.11.1.   Is the technical feasibility project being complied with?

4.12.   Do you use agricultural credit? For funding or investment? How much? Do you pay annually?

5.   Commercialisation.

5.1.   How is production commercialised? Is it done directly with the "market" or is there a middleman?

5.2.   Do you have your own transport?

5.3.   What is the proportional share of transport in the value obtained from the sale of production?

5.4.   Do you think production prices are getting better or worse?

5.5.   Is there an increase in the placement of your production on the market?

I want morebooks!

Buy your books fast and straightforward online - at one of world's fastest growing online book stores! Environmentally sound due to Print-on-Demand technologies.

Buy your books online at
www.morebooks.shop

Kaufen Sie Ihre Bücher schnell und unkompliziert online – auf einer der am schnellsten wachsenden Buchhandelsplattformen weltweit! Dank Print-On-Demand umwelt- und ressourcenschonend produziert.

Bücher schneller online kaufen
www.morebooks.shop

info@omniscriptum.com
www.omniscriptum.com

Printed by Books on Demand GmbH, Norderstedt / Germany